Maths Skills *for A Level*

Chemistry

2nd Edition

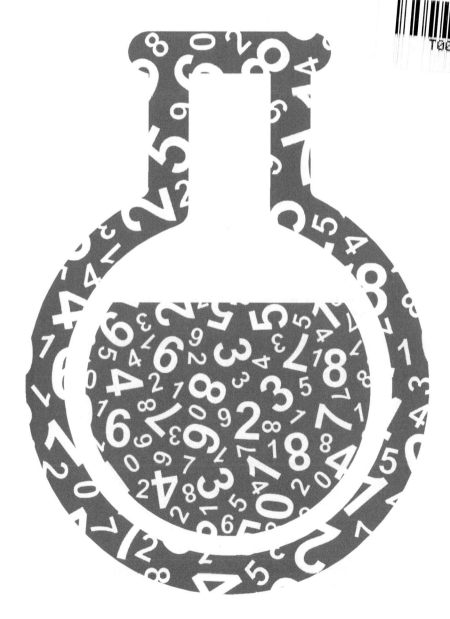

Emma Poole and Dan McGowan

OXFORD

UNIVERSITY PRESS

OXFORD
UNIVERSITY PRESS

Great Clarendon Street, Oxford, OX2 6DP, United Kingdom

Oxford University Press is a department of the University of Oxford.
It furthers the University's objective of excellence in research, scholarship,
and education by publishing worldwide. Oxford is a registered trade mark of
Oxford University Press in the UK and in certain other countries

British Library Cataloguing in Publication Data
Data available

978-0-19-842897-8

10 9 8 7 6 5 4

Printed and bound by CPI Group (UK) Ltd, Croydon, CR0 4YY

Acknowledgements

Cover: OUP/Shutterstock

Artwork by Thomson Digital and Tech-Set Ltd, Gateshead

In-house editor Guy Williamson

Although we have made every effort to trace and contact all
copyright holders before publication this has not been possible in all
cases. If notified, the publisher will rectify any errors or omissions at
the earliest opportunity.

Links to third party websites are provided by Oxford in good faith
and for information only. Oxford disclaims any responsibility for
the materials contained in any third party website referenced in
this work.

Contents

How to use this book

This workbook has been written to support the development of key mathematics skills required to achieve success in your A Level Chemistry course. It has been devised and written by teachers and the practice questions included reflect the **exam-tested content** for AQA, OCR, and Cambridge specifications.

The workbook is structured into chapters with each chapter having a clear scientific topic. Then, each topic covers a mathematical skill or skills that you may need to practise. Each topic offers the following features:

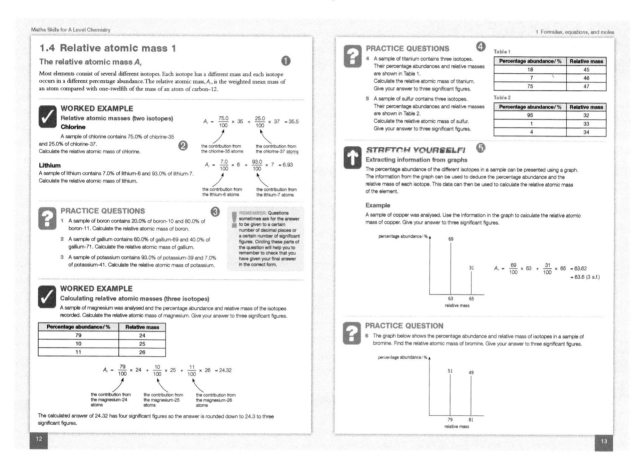

① **Opening paragraph** outlines the mathematical skill or skills covered within the topic.

② **Worked example** – each topic will have one or two worked examples. The worked examples will be annotated.

③ **Remember** is a useful box that will offer you tips, hints, and other snippets of useful information.

④ **Practice questions** are ramped in terms of difficulty and all answers are available at www.oxfordsecondary.co.uk.

⑤ **Stretch yourself** – some of the topics may also contain a few more difficult questions and concepts to stretch your mathematical knowledge and understanding.

1 FORMULAE, EQUATIONS, AND MOLES

1.1 Balancing equations and rearranging equations

Conservation of mass

New substances are made during chemical reactions. However, atoms are not created or destroyed: they just become rearranged in new ways. This means that there is always the same number of each type of atom before and after the reaction, so the total mass before the reaction must be the same as the total mass after the reaction. This is known as the conservation of mass.

WORKED EXAMPLE

Balancing an equation

Balance the equation below.

$H_2 + O_2 \rightarrow H_2O$

The unbalanced equation suggests that one hydrogen molecule, consisting of two hydrogen atoms, reacts with one oxygen molecule, consisting of two oxygen atoms, to form one water molecule, consisting of two hydrogen atoms and one oxygen atom.

The equation shows the correct formulae but it is not balanced.

There are two hydrogen atoms on both sides of the equation but whilst there are two oxygen atoms on the left-hand side of the equation there is only one oxygen atom on the right-hand side of the equation. Therefore, a two must be placed before the H_2O.

$H_2 + O_2 \rightarrow 2H_2O$

Now the oxygen atoms are balanced but the hydrogen atoms are no longer balanced. A two must be placed in front of the H_2.

$2H_2 + O_2 \rightarrow 2H_2O$

The number of hydrogen and oxygen atoms is the same on both sides, so the equation is balanced.

PRACTICE QUESTION

1 Balance the following equations.

    a $H_2 + I_2 \rightarrow HI$

    b $C + O_2 \rightarrow CO$

    c $N_2 + H_2 \rightarrow NH_3$

    d $H_2SO_4 + KOH \rightarrow K_2SO_4 + H_2O$

    e $C_2H_4 + O_2 \rightarrow H_2O + CO_2$

WORKED EXAMPLE

Balancing an equation with fractions

Balance the equation below.

$C_2H_6 + O_2 \rightarrow CO_2 + H_2O$

The number of carbon atoms is balanced by placing a two before the CO_2, whilst the hydrogen atoms are balanced by placing a three in front of the H_2O.

$C_2H_6 + O_2 \rightarrow 2CO_2 + 3H_2O$

This means that there are four oxygen atoms in the carbon dioxide molecules plus three oxygen atoms in the water molecules, giving a total of seven oxygen atoms on the product side. To balance the equation, three and a half must be placed in front of the O_2.

$C_2H_6 + 3\tfrac{1}{2}O_2 \rightarrow 2CO_2 + 3H_2O$

If desired, the equation can be multiplied by 2 to get whole numbers.

$2C_2H_6 + 7O_2 \rightarrow 4CO_2 + 6H_2O$

PRACTICE QUESTION

2 Balance the equations below.

    a $C_3H_6 + O_2 \rightarrow CO_2 + H_2O$

    b $C_6H_{14} + O_2 \rightarrow CO_2 + H_2O$

    c $NH_2CH_2COOH + O_2 \rightarrow CO_2 + H_2O + N_2$

WORKED EXAMPLE

Balancing an equation with brackets

Balance the equation below.

$Ca(OH)_2 + HCl \rightarrow CaCl_2 + H_2O$

Here the brackets around the hydroxide (OH^-) group show that the $Ca(OH)_2$ unit contains one calcium atom, two oxygen atoms, and two hydrogen atoms. To balance the equation, a 2 must be placed before the HCl and another before the H_2O.

$Ca(OH)_2 + 2HCl \rightarrow CaCl_2 + 2H_2O$

The equation is now balanced.

PRACTICE QUESTION

3 Balance the equations below.

    a $Mg(OH)_2 + HNO_3 \rightarrow Mg(NO_3)_2 + H_2O$

    b $Fe(NO_3)_2 + Na_3PO_4 \rightarrow Fe_3(PO_4)_2 + NaNO_3$

WORKED EXAMPLE

Rearranging expressions

In chemistry you sometimes need to rearrange an expression to find the desired values.

You may know the amount of a substance (n) and the mass of it you have (m) and need to find its molar mass (M). The amount of substance (n) is equal to the mass you have (m) divided by the molar mass (M).

$$n = \frac{m}{M}$$

You need to rearrange the expression to make the molar mass (M) the subject.

Multiply both sides by the molar mass (M)

$$M \times n = m$$

Then divide both sides by the amount of substance (n).

$$M = \frac{m}{n}$$

PRACTICE QUESTIONS

4 Rearrange the expression $c = \dfrac{n}{v}$

    to make:

    a n the subject of the expression

    b v the subject of the expression.

5 Rearrange the expression $PV = nRT$

    to make:

    a n the subject of the expression

    b P the subject of the expression

    c T the subject of the expression.

1.2 Amount of substance

A mole of substance

Atoms are too small to be counted, so chemists measure the amount of substance using moles. One mole of any substance contains the same number of particles as there are carbon atoms in 12.0 g of carbon-12.

WORKED EXAMPLE

Mass of one mole of a substance

The mass of one mole of any element is its relative atomic mass in grams.
One mole of carbon has a mass of 12.0 g.

One mole of sulfur has a mass of 32.1 g.

One mole of copper has a mass of 63.5 g.

PRACTICE QUESTION

1 Calculate the mass, in g, of:

    a 1.00 mole of magnesium

    b 1.00 mole of beryllium

    c 1.00 mole of aluminium.

WORKED EXAMPLE

Mass of different amounts of substance

To work out the mass of other amounts of substance, in moles, multiply the relative atomic mass in grams by the amount of substance required in moles.

2.00 moles of carbon has a mass of 12.0 g × 2 = 24.0 g

3.00 moles of sulfur has a mass of 32.1 g × 3 = 96.3 g

2.50 moles of copper has a mass of 63.5 g × 2.5 = 158.75 g

> **REMEMBER:** The molar mass of a substance is the mass per mole of the substance. The units are $g \, mol^{-1}$.

PRACTICE QUESTION

2 Calculate the mass, in g, of:

    a 2.00 moles of magnesium

    b 1.50 moles of beryllium

    c 2.00 moles of aluminium

    d 5.00 moles of oxygen

    e 1.50 moles of neon.

WORKED EXAMPLE

The molar mass of carbon, C, is simply the relative atomic mass of carbon, which is 12.0 in units of $g \, mol^{-1}$. So, the molar mass of carbon is $12.0 \, g \, mol^{-1}$.

The molar mass of water, H_2O, is the total of the relative atomic masses of all the atoms that make up a relative formula unit of H_2O.

$(2 \times 1.0) + (1 \times 16.0) = 18.0 \, g \, mol^{-1}$

PRACTICE QUESTION

3 Calculate the molar mass of:

    a sulfur, S

    b sulfur dioxide, SO_2

    c carbon dioxide, CO_2

    d boron trichloride, BCl_3

    e ammonia, NH_3.

WORKED EXAMPLE

Linking moles, mass, and molar mass

The number of moles (n), the mass of the substance (m) and the molar mass (M) are linked together using $n = \dfrac{m}{M}$.

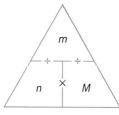

n = number of moles, in mol
m = mass of substance, in g
M = molar mass, in $g\,mol^{-1}$

1 Calculate the number of moles in 6.0 g of carbon.

$$n = \frac{m}{M}$$

$$= \frac{6.0\,g}{12.0\,g\,mol^{-1}}$$

$$= 0.5\,mol$$

2 0.5 moles of a substance has a mass of 22 g. Calculate the molar mass of the compound.

$$M = \frac{m}{n}$$

$$= \frac{22.0\,g}{0.5\,mol}$$

$$= 44.0\,g\,mol^{-1}$$

PRACTICE QUESTIONS

4 Calculate the amount, in moles, of:

    a 9.0 g of carbon, C

    b 36.45 g of magnesium, Mg

    c 76.0 g of fluorine, F_2.

5 Calculate the molar mass of a compound when:

    a 0.25 moles of a compound has a mass of 25.0 g

    b 0.10 moles of a compound has a mass of 4.4 g

    c 0.05 moles of a compound has a mass of 5.0 g.

1.3 Avogadro constant

The amount of substance

Chemists measure the amount of substance in moles. There is the same number of particles in one mole of any substance. This number of particles is known as the Avogadro constant, N_A. The Avogadro constant is a large number and is normally written in standard form as $6.02 \times 10^{23} \, mol^{-1}$ (to three significant figures).

WORKED EXAMPLE

Finding the number of particles in one mole

1 Calculate the number of atoms in 1.00 mole of carbon.
1.00 mole of carbon contains 6.02×10^{23} atoms.

2 Calculate the number of atoms in 1.00 mole of sodium.
1.00 mole of sodium contains 6.02×10^{23} atoms.

PRACTICE QUESTION

1 Calculate the number of atoms in:

   a 1.00 mole of lithium

   b 1.00 mole of tungsten

   c 1.00 mole of aluminium.   Give your answers to three significant figures.

WORKED EXAMPLE

Finding the number of atoms in different amounts of substance

To work out the number of atoms in other amounts of a substance, multiply the number of moles (n) by the Avogadro constant (N_A). L can also be used for the Avogadro constant.

Number of atoms $= n \times N_A$

Calculate the number of atoms in 2.00 moles of magnesium atoms.

2.00 moles of magnesium atoms $= 2.00 \times 6.02 \times 10^{23} = 1.204 \times 10^{24}$ atoms

Notice that when you write the number in standard form the power is now $\times 10^{24}$.

PRACTICE QUESTIONS

2 Calculate the number of atoms in:

   a 0.10 moles of carbon   **b** 2.50 moles of sulfur   **c** 0.75 moles of magnesium.

3 Calculate the number of moles of atoms in:

   a 2.50 moles of sulfur (S_8)   **b** 3.00 moles of butane (C_4H_{10}).

   Give your answers to three significant figures.

Other types of particle

The Avogadro constant can be used to work out the number of particles in any type of substance.

PRACTICE QUESTION

4 Calculate the number of particles in:

   a 1.00 mole of sodium ions   **b** 1.00 mole of nitrogen molecules

   c 1.00 mole of magnesium ions.   Give your answers to three significant figures.

WORKED EXAMPLE

Finding the number of particles in different amounts of substance

To work out the numbers of particles in other amounts of a substance, multiply the amount of substance, in moles, by the Avogadro constant.

Number of particles $= n \times N_A$

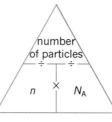

Calculate the number of ions in 3.00 moles of aluminium ions.

Number of particles (ions) $= 3.00 \, \text{mol} \times 6.02 \times 10^{23} \, \text{mol}^{-1} = 1.806 \times 10^{24}$ ions

PRACTICE QUESTION

5   Calculate the number of particles in:

    a   2.00 moles of electrons

    b   1.50 moles of oxide ions

    c   0.2 moles of lithium ions.

STRETCH YOURSELF!

Linking the Avogadro constant to the number of grams of a substance

The number of particles in a given mass of a substance can be calculated using a two-step calculation.

First work out the number of moles using $n = \dfrac{m}{M}$

where n is the number of moles, m is the mass of the substance, and M is the molar mass of the substance.

Next find the number of particles using:

   number of particles $= n \times N_A$

Example

Calculate the number of particles in 18.0 g of carbon.

$$n = \frac{m}{M}$$

$$= \frac{18.0 \, \text{g}}{12.0 \, \text{g mol}^{-1}}$$

$$= 1.50 \, \text{mol}$$

Number of particles $= n \times N_A$

$$= 1.50 \times 6.02 \times 10^{23} = 9.03 \times 10^{23}$$

PRACTICE QUESTION

6   Calculate the number of particles in 48.6 g of magnesium atoms.

1.4 Relative atomic mass 1

The relative atomic mass A_r

Most elements consist of several different isotopes. Each isotope has a different mass and each isotope occurs in a different percentage abundance. The relative atomic mass, A_r, is the weighted mean mass of an atom compared with one-twelfth of the mass of an atom of carbon–12.

WORKED EXAMPLE

Relative atomic masses (two isotopes)
Chlorine

A sample of chlorine contains 75.0% of chlorine-35 and 25.0% of chlorine-37.
Calculate the relative atomic mass of chlorine.

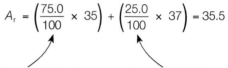

$$A_r = \left(\frac{75.0}{100} \times 35\right) + \left(\frac{25.0}{100} \times 37\right) = 35.5$$

the contribution from the chlorine-35 atoms

the contribution from the chlorine-37 atoms

Lithium

A sample of lithium contains 7.0% of lithium-6 and 93.0% of lithium-7.
Calculate the relative atomic mass of lithium.

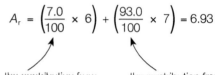

$$A_r = \left(\frac{7.0}{100} \times 6\right) + \left(\frac{93.0}{100} \times 7\right) = 6.93$$

the contribution from the lithium-6 atoms

the contribution from the lithium-7 atoms

PRACTICE QUESTIONS

1   A sample of boron contains 20.0% of boron-10 and 80.0% of boron-11. Calculate the relative atomic mass of boron.

2   A sample of gallium contains 60.0% of gallium-69 and 40.0% of gallium-71. Calculate the relative atomic mass of gallium.

3   A sample of potassium contains 93.0% of potassium-39 and 7.0% of potassium-41. Calculate the relative atomic mass of potassium.

> **REMEMBER:** Questions sometimes ask for the answer to be given to a certain number of decimal places or a certain number of significant figures. Circling these parts of the question will help you to remember to check that you have given your final answer in the correct form.

WORKED EXAMPLE

Calculating relative atomic masses (three isotopes)

A sample of magnesium was analysed and the percentage abundance and relative mass of the isotopes recorded. Calculate the relative atomic mass of magnesium. Give your answer to three significant figures.

Percentage abundance / %	Relative mass
79	24
10	25
11	26

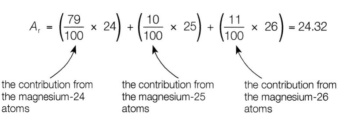

$$A_r = \left(\frac{79}{100} \times 24\right) + \left(\frac{10}{100} \times 25\right) + \left(\frac{11}{100} \times 26\right) = 24.32$$

the contribution from the magnesium-24 atoms

the contribution from the magnesium-25 atoms

the contribution from the magnesium-26 atoms

The calculated answer of 24.32 has four significant figures so the answer is rounded down to 24.3 to three significant figures.

PRACTICE QUESTIONS

4 A sample of titanium contains three isotopes. Their percentage abundances and relative masses are shown in Table 1.
Calculate the relative atomic mass of titanium.
Give your answer to three significant figures.

5 A sample of sulfur contains three isotopes. Their percentage abundances and relative masses are shown in Table 2.
Calculate the relative atomic mass of sulfur.
Give your answer to three significant figures.

Table 1

Percentage abundance / %	Relative mass
18	45
7	46
75	47

Table 2

Percentage abundance / %	Relative mass
95	32
1	33
4	34

STRETCH YOURSELF!

Extracting information from graphs

The percentage abundance of the different isotopes in a sample can be presented using a graph. The information from the graph can be used to deduce the percentage abundance and the relative mass of each isotope. This data can then be used to calculate the relative atomic mass of the element.

Example

A sample of copper was analysed. Use the information in the graph to calculate the relative atomic mass of copper. Give your answer to three significant figures.

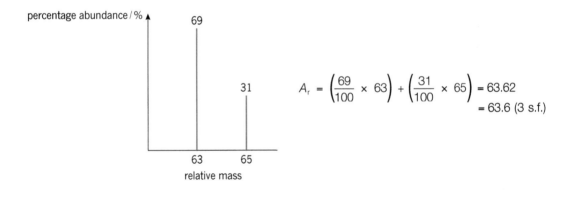

$$A_r = \left(\frac{69}{100} \times 63\right) + \left(\frac{31}{100} \times 65\right) = 63.62$$
$$= 63.6 \text{ (3 s.f.)}$$

PRACTICE QUESTION

6 The graph below shows the percentage abundance and relative mass of isotopes in a sample of bromine. Find the relative atomic mass of bromine. Give your answer to three significant figures.

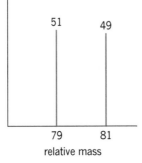

1.5 Relative atomic mass 2

Isotopes

Isotopes have the same atomic number but a different mass number. This means they have the same number of protons but a different number of neutrons.

In atoms that are neutral, the number of protons is equal to the number of electrons. Subatomic units are the protons, neutrons, and electrons that are found in atoms.

WORKED EXAMPLES

Subatomic units

A sodium atom has the symbol $_{11}^{23}\text{Na}$

Calculate the number of protons, neutrons, and electrons in this atom of sodium-23.
Number of protons = atomic number = 11
Number of neutrons = mass number − atomic number = 23 − 11 = 12
Number of electrons = protons = 11

A fluoride ion has the symbol $_{9}^{19}\text{F}^-$

Calculate the number of protons, neutrons, and electrons in this ion of fluorine-19.
Number of protons = atomic number = 9
Number of neutrons = mass number − atomic number = 19 − 9 = 10
Number of electrons = number of protons plus an extra one as it has a 1− charge = 9 + 1 = 10

PRACTICE QUESTIONS

1 Calculate the number of protons, neutrons, and electrons in an atom of magnesium-25.

2 Calculate the number of protons, neutrons, and electrons in an atom of lithium-7.

3 Calculate the number of protons, neutrons, and electrons in an ion of chlorine-35 with a 1− charge.

4 Calculate the number of protons, neutrons, and electrons in an ion of potassium-39 with a 1+ charge.

5 Calculate the number of protons, neutrons, and electrons in an ion of calcium-40 with a 2+ charge.

WORKED EXAMPLE

Using relative atomic mass to calculate the percentage abundances of isotopes

Lithium has two stable isotopes: lithium-6 and lithium-7. Lithium has a relative atomic mass of 6.92.
Calculate the relative atomic abundance of the two isotopes. Give your answer to two significant figures.
The sum of the percentage abundances of lithium-6 and lithium-7 must be equal to 100%
If the percentage abundance of lithium-6 = x

The percentage abundance of lithium-7 = $(100 − x)$
If there were 100 atoms, x would be lithium-6 and $(100 − x)$ would be lithium-7.
$(x \times 6) + (100 − x) \times 7 = 6.92 \times 100$
$6x + 700 − 7x = 692$
$-x = −8$
$x = 8$

So the percentage abundance of lithium-6 = x = 8.0 (2 s.f.)
The percentage abundance of lithium-7 = $(100 − x)$ = 92 (2 s.f.)

PRACTICE QUESTIONS

6 Bromine has two isotopes: bromine-79 and bromine-81.

It has relative atomic mass of 79.99.

Calculate the percentage abundance of these two isotopes. Give your answer to three significant figures.

7 Copper has two isotopes: copper-63 and copper-65.

It has relative atomic mass of 63.5.

Calculate the percentage abundance of these two isotopes. Give your answer to three significant figures.

Predicting the mass spectra of diatomic molecules

A chlorine molecule has the formula Cl_2.

Chlorine has two isotopes: chlorine-35 and chlorine-37.

75% or $\frac{3}{4}$ of chlorine atoms have a relative atomic mass of 35.

25% or $\frac{1}{4}$ of chlorine atoms have a relative atomic mass of 37.

In a chlorine molecule there are three possible combinations of chlorine atoms:

1 $^{35}Cl-^{35}Cl$ **2** $^{37}Cl-^{37}Cl$ **3** $^{37}Cl-^{35}Cl$ or $^{35}Cl-^{37}Cl$

If chlorine was placed in a mass spectrometer:

The first combination of atoms would give a peak at $m/z = 70$ (35 + 35)

The second combination of atoms would give a peak at $m/z = 74$ (37 + 37)

The third combination of atoms would give a peak at $m/z = 72$ (37 + 35 or 35 + 37)

The chance of producing a peak at $m/z = 70 = \frac{3}{4} \times \frac{3}{4} = \frac{9}{16}$

The chance of producing a peak at $m/z = 74 = \frac{1}{4} \times \frac{1}{4} = \frac{1}{16}$

The chance of producing a peak at $m/z = 72$

$$= \left(\frac{1}{4} \times \frac{3}{4}\right) + \left(\frac{3}{4} \times \frac{1}{4}\right) = \left(\frac{3}{16} + \frac{3}{16}\right) = \frac{6}{16}$$

So the ratio of the heights of the peaks at $70 : 74 : 72$ will be $9 : 1 : 6$.

PRACTICE QUESTION

8 A bromine molecule has the formula Br_2.

A sample of bromine has two isotopes: bromine-79 and bromine-81.

50% or $\frac{1}{2}$ of bromine atoms have a relative atomic mass of 79.

50% or $\frac{1}{2}$ of bromine atoms have a relative atomic mass of 81.

Predict the mass spectrum produced by a sample of bromine molecules.

1.6 Moles and gas volumes

Avogadro's law

Avogadro's law states that equal volumes of gases at the same temperature and pressure contain an equal number of moles. At room temperature and pressure (RTP), 1 mole of any gas takes up a volume of $24\,dm^3$ or $24\,000\,cm^3$. Room temperature and pressure are taken to be 25 °C and 1 atmosphere. This means that at room temperature and pressure, 1 mole of helium atoms takes up a volume of $24\,dm^3$ or $24\,000\,cm^3$.

At room temperature and pressure 1 mole of oxygen molecules also takes up a volume of $24\,dm^3$ or $24\,000\,cm^3$. Molar gas volume has units of $dm\,mol^{-1}$.

WORKED EXAMPLES

Calculations using dm^3

Amount of substance from volume of gas at RTP

The amount of substance (n), in moles, and the volume (V) of gas, in dm^3 are linked together by the equation:

$$n = \frac{V \text{ (in } dm^3)}{24.0}$$

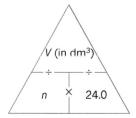

Number of moles

Calculate the number of moles of gas in $12\,dm^3$ of nitrogen, N_2, at RTP.

$$n = \frac{V \text{ (in } dm^3)}{24.0}$$

$$n = \frac{12}{24.0} = 0.5\,mol$$

PRACTICE QUESTION

1 Calculate the number of moles of gas at RTP in:

a $6\,dm^3$ of oxygen, O_2 b $36\,dm^3$ of carbon dioxide, CO_2

c $120\,dm^3$ of water vapour, H_2O d $72\,dm^3$ of carbon dioxide, CO_2

e $12\,dm^3$ of water vapour, H_2O.

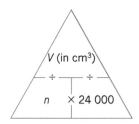

WORKED EXAMPLES

Calculations using cm^3

Amount of substance from the volume of gas at RTP

The amount of substance (n), in moles, and the volume (V) of gas in cm^3 are linked together by the equation:

$$n = \frac{V \text{ (in } cm^3)}{24\,000}$$

Number of moles

Calculate the number of moles of gas in $600\,cm^3$ of oxygen, O_2, at RTP.

$$n = \frac{V \text{ (in } cm^3)}{24\,000}$$

$$n = \frac{600}{24\,000} = 0.025\,mol$$

PRACTICE QUESTION

2   Calculate the number of moles of gas at RTP in:

   a $8000\,cm^3$ of carbon dioxide, CO_2 **b** $72\,000\,cm^3$ of carbon monoxide, CO

   c $1800\,dm^3$ of sulfur dioxide, SO_2 **d** $12\,000\,cm^3$ of carbon monoxide, CO

   e $3600\,dm^3$ of sulfur dioxide, SO_2.

WORKED EXAMPLES

Calculating the volume in dm^3 that an amount of gas occupies at RTP

The volume of gas can be calculated using

$$V \text{ (in dm}^3) = n \times 24.0$$

Calculate the volume, in dm^3, of 2.75 moles of a gas at RTP.

$$V \text{ (in dm}^3) = 2.75 \times 24.0 = 66\,dm^3$$

PRACTICE QUESTION

3   Calculate the volume, in dm^3, of these gases at RTP:

   a   0.10 moles of carbon dioxide, CO_2 **b**   2.50 moles of sulfur dioxide, SO_2

   c   0.20 moles of water vapour, H_2O **d**   15.0 moles of sulfur dioxide, SO_2

   e   0.05 moles of water vapour, H_2O.

STRETCH YOURSELF!

Linking the volume of gas to its mass

The mass of a gas in a given volume can be calculated using a two-step calculation.

First work out the number of moles using $n = \dfrac{V \text{ (in dm}^3)}{24.0}$

Next work out the mass of the gas using:

$$m = n \times M$$

Example

Calculate the mass of $0.60\,dm^3$ of carbon dioxide, CO_2.

$$n = \frac{V \text{ (in dm}^3)}{24.0}$$

$$n = \frac{0.60}{24.0} = 0.025\,mol$$

$$m = n \times M$$

$$m = 0.025\,mol \times 44\,g\,mol^{-1} = 1.1\,g$$

PRACTICE QUESTION

4   **a**   Calculate the mass of $2.00\,dm^3$ of sulfur dioxide, SO_2.

   b   Calculate the mass of $12.00\,dm^3$ of carbon dioxide, CO_2.

   c   Calculate the mass of $120.00\,dm^3$ of oxygen, O_2.

1.7 Ideal gas equation

$pV = nRT$

The ideal gas equation links:

- the pressure (p) in pascal, Pa
- the amount of gas (n) in mol
- the temperature (T) in kelvin, K.
- the volume (V) in m^3
- the gas constant (R), which is equal to $8.31\,\text{J}\,\text{K}^{-1}\,\text{mol}^{-1}$

The ideal gas equation is $pV = nRT$

WORKED EXAMPLE

Unit conversions for temperature

The Kelvin scale is used to measure the absolute temperature. It has no negative values. Absolute zero, 0 K, is equal to $-273\,°\text{C}$.

   0 °C is equal to 273 K.

Convert 25 °C into kelvin.

   25 °C = 273 + 25 = 298 K

PRACTICE QUESTION

1 Convert the following temperatures into kelvin:

   a 100 °C   b 37 °C   c −100 °C   d 200 °C   e −200 °C.

WORKED EXAMPLE

Unit conversions for volume

The ideal gas equation requires that the volume must be measured in m^3.

   1 dm^3 is equal to 1×10^{-3} m^3 or 0.001 m^3.

Convert 2 dm^3 into m^3.

   2 dm^3 = $2 \times 1 \times 10^{-3}$ = 2×10^{-3} m^3

PRACTICE QUESTION

2 Convert the following volumes into m^3:

   a 0.1 dm^3 b 10 dm^3 c 150 dm^3 d 0.5 dm^3 e 20 dm^3.

WORKED EXAMPLE

Unit conversions for pressure

Finally, the ideal gas equation requires that the pressures must be measured in pascals, Pa.

   1 kilopascal, kPa, is equal to 1×10^3 Pa or 1000 Pa.

Convert 5 kPa into Pa.

   5 kPa = $5 \times 1 \times 10^3$ = 5×10^3 Pa or 5000 Pa

PRACTICE QUESTION

3 Convert the following pressures into pascals, Pa:

   a 0.1 kPa   b 10 kPa   c 2.5 kPa   d 100 kPa   e 0.5 kPa.

Finding a volume

The volume of a gas can be calculated using $V = \dfrac{nRT}{p}$.

WORKED EXAMPLE

Calculate the volume of 5.0×10^{-3} moles of oxygen at a pressure of 10 Pa at 0 °C.
Give your answer to three significant figures.

0 °C = 273 K

$$V = \frac{5.0 \times 10^{-3} \times 8.31 \times 273}{10} = 1.13\,m^3 \text{ (3 s.f.)}$$

PRACTICE QUESTION

4   Calculate the volume of 5.0×10^{-6} moles of hydrogen at a pressure of 100 Pa at 25 °C.
    Give your answer to three significant figures.

Finding a pressure

The pressure of a gas can be calculated using $p = \dfrac{nRT}{V}$.

WORKED EXAMPLE

Calculate the pressure exerted by 5.0 moles of hydrogen at 25 °C in a 10 m³ container.
Give your answer to three significant figures.

$$p = \frac{5.0 \times 8.31 \times 298}{10} = 1240\,Pa \text{ (3 s.f.)}$$

PRACTICE QUESTION

5   Calculate the pressure exerted by 0.2 moles of carbon dioxide in a 10 m³ container at 0 °C.
    Give your answer to three significant figures.

Finding a temperature

The temperature of a gas can be calculated using $T = \dfrac{pV}{nR}$.

WORKED EXAMPLE

Calculate the temperature of 5.0 moles of carbon dioxide at 500 Pa in a 20 dm³ container.
Give your answer to three significant figures.

20 dm³ = 20 × 1 × 10⁻³ = 0.02 m³

$$T = \frac{500 \times 0.02}{5.0 \times 8.31} = 0.241\,K \text{ (3 s.f.)}$$

PRACTICE QUESTION

6   Calculate the temperature of 0.5 moles of oxygen at 100 Pa in a 250 m³ container. Give your answer to three significant figures.

REMEMBER: You may need to convert several of the pieces of information given to you into the units required in the ideal gas equation.

1.8 Concentration

Making a solution

A solution is made when a solute dissolves in a solvent. The concentration of a solution is a way of saying how much solute, in moles, is dissolved in $1\,dm^3$ or 1 litre of solution. Concentration is usually measured using units of $mol\,dm^{-3}$.

The concentration of the amount of substance dissolved in a given volume of a solution is given by the equation:

$$c = \frac{n}{V\ (\text{in dm}^3)}$$

Where n is the amount of substance in moles, c is the concentration, and V is the volume in dm^3. Concentration can also be measured in $g\,dm^{-3}$.

WORKED EXAMPLE

Calculating concentration

1 Calculate the concentration in $mol\,dm^{-3}$ of a solution formed when 0.5 moles of a solute is dissolved in $2\,dm^3$ of solution.

$$c = \frac{n}{V(\text{in dm}^3)} = \frac{0.5\,mol}{2\,dm^3} = 0.25\,mol\,dm^{-3}$$

2 Calculate the concentration in $mol\,dm^{-3}$ of a solution formed when 0.1 moles of a solute is dissolved in $500\,cm^3$ of solution.

$500\,cm^3 = 0.5\,dm^3$

$$c = \frac{n}{V\ (\text{in dm}^3)} = \frac{0.1\,mol}{0.5\,dm^3} = 0.2\,mol\,dm^{-3}$$

PRACTICE QUESTIONS

1 Calculate the concentration, in $mol\,dm^{-3}$, of a solution formed when 0.2 moles of a solute is dissolved in $50\,cm^3$ of solution.

2 Calculate the concentration, in $mol\,dm^{-3}$, of a solution formed when 0.25 moles of a solute is dissolved in $0.1\,dm^3$ of solution.

3 Calculate the concentration, in $mol\,dm^{-3}$, of a solution formed when 0.05 moles of a solute is dissolved in $2.0\,dm^3$ of solution.

Calculating the amount of substance from the concentration and volume of the solution

The equation can be rearranged to calculate the amount of substance, in moles, from a known volume and concentration of solution.

$$n = c \times V\ (\text{in dm}^3)$$

WORKED EXAMPLE

Calculate the number of moles in a solution of sodium hydroxide, $NaOH$, in $25\,cm^3$ of aqueous solution of concentration $0.5\,mol\,dm^{-3}$.

$n = c \times V\ (\text{in dm}^3)$

$n = 0.5\,mol\,dm^{-3} \times 0.025\,dm^3 = 0.0125\,mol$

PRACTICE QUESTION

4 Calculate the number of moles of NaOH in an aqueous solution of:

    a $36\,cm^3$ of $0.1\,mol\,dm^{-3}$ **b** $26\,cm^3$ of $0.5\,mol\,dm^{-3}$

    c $50\,cm^3$ of $0.05\,mol\,dm^{-3}$.

Calculating the volume of a solution from a given amount of substance and concentration

The equation can be rearranged to calculate the volume of a solution from a known amount of substance, in moles, and the concentration of the solution.

$$V \text{ (in dm}^3) = \frac{n}{c}$$

WORKED EXAMPLE

Calculate the volume, in dm^3, of a solution of concentration $0.5\,mol\,dm^{-3}$ that contains 0.05 moles of the solute.

$$V \text{ (in dm}^3) = \frac{n}{c} = \frac{0.05\,mol}{0.5\,mol\,dm^{-3}} = 0.1\,dm^3$$

PRACTICE QUESTIONS

5 Calculate the volume, in dm^3, of a solution of concentration $0.10\,mol\,dm^{-3}$ that contains 0.01 moles of the solute.

6 Calculate the volume, in dm^3, of a solution of concentration $0.05\,mol\,dm^{-3}$ that contains 0.25 moles of the solute.

STRETCH YOURSELF!

Linking the mass of a solute to the volume and concentration of a solution

The mass of a solute in a solution of a given volume and concentration can be calculated using a two-step calculation.

First work out the number of moles using $n = c \times V$ (in dm^3).

Next work out the mass of the solute using:

    $m = n \times M$

where m is the mass of the substance, n is the amount of substance in moles, and M is the molar mass.

Find the mass of sodium hydroxide, NaOH, required to prepare $250\,cm^3$ of an aqueous solution with a concentration of $0.10\,mol\,dm^{-3}$.

    $n = c \times V$ (in dm^3)

    $n = 0.1\,mol\,dm^{-3} \times 0.25 = 0.025\,mol$

    $m = n \times M$

    $m = 0.025\,mol \times 40\,g\,mol^{-1} = 1\,g$

PRACTICE QUESTION

7 Find the mass of sodium hydroxide, NaOH, required to prepare $100\,cm^3$ of an aqueous solution with a concentration of $0.20\,mol\,dm^{-3}$.

1.9 Titrations

Acid–base titrations

In acid–base titrations, a solution of an acid reacts with a solution of a base. The concentration of one of the solutions is known. A titration can be used to work out the concentration of the other solution. The amount of substance (n) in moles, the volume of solution (V) in cm^3, and the concentration (c) are linked together by the equation:

$$n = \frac{V \text{ (in } cm^3)}{1000} \times c$$

WORKED EXAMPLE

Finding the concentration of a solution

Caeli carries out a titration to find the concentration of some hydrochloric acid. She finds that $22.50\,cm^3$ of hydrochloric acid was required to neutralise $25.00\,cm^3$ of $0.10\,mol\,dm^{-3}$ aqueous sodium hydroxide solution.

$$HCl(aq) + NaOH(aq) \rightarrow NaCl(aq) + H_2O(l)$$

a Calculate the number of moles of sodium hydroxide used.

$$n = \frac{V \text{ (in } cm^3)}{1000} \times c = \frac{25.00}{1000} \times 0.1 = 0.0025 \text{ mol}$$

b Calculate the number of moles of hydrochloric acid used.

From the stoichiometric equation, the number of moles of hydrochloric acid used = 0.0025 mol.

c Calculate the concentration of the hydrochloric acid. Give your answer to two decimal places.

$$c = \frac{n \times 1000}{V} = \frac{0.0025 \times 1000}{22.50} = 0.11\,mol\,dm^{-3}$$

PRACTICE QUESTIONS

1 Roweena carries out a titration to find the concentration of some nitric acid. Roweena finds that $50\,cm^3$ of $0.125\,mol\,dm^{-3}$ aqueous sodium hydroxide solution was neutralised by $22.50\,cm^3$ of the nitric acid. Give your answers to two decimal places.

$$HNO_3(aq) + NaOH(aq) \rightarrow NaNO_3(aq) + H_2O(l)$$

  a Calculate the number of moles of sodium hydroxide used.
  b Calculate the number of moles of nitric acid used.
  c Calculate the concentration of the nitric acid.

2 Shay carries out a titration to find the concentration of some hydrochloric acid. $25.0\,cm^3$ of a standard solution of $0.20\,mol\,dm^{-3}$ sodium hydroxide was placed into a conical flask. Shay found that $22.0\,cm^3$ of hydrochloric acid was required for neutralisation.

$$HCl(aq) + NaOH(aq) \rightarrow NaCl(aq) + H_2O(l)$$

  a Calculate the number of moles of sodium hydroxide used.
  b Calculate the number of moles of hydrochloric acid used.
  c Calculate the concentration of the hydrochloric acid.

WORKED EXAMPLE

Harrison carries out a titration to find the concentration of some sodium hydroxide solution. She finds that $25.00\,cm^3$ of aqueous sodium hydroxide solution was neutralised by $28.00\,cm^3$ of $0.08\,mol\,dm^{-3}$ sulfuric acid.

$$H_2SO_4(aq) + 2NaOH(aq) \rightarrow Na_2SO_4(aq) + 2H_2O(l)$$

a Calculate the number of moles of sulfuric acid used.

$$n = \frac{V \text{ (in cm}^3)}{1000} \times c = \frac{28.00}{1000} \times 0.08 = 0.00224 \text{ mol}$$

b Calculate the number of moles of sodium hydroxide used.

From the stoichiometric equation, the number of moles of sodium hydroxide used = 0.00448 mol.

c Calculate the concentration of the sodium hydroxide. Give your answer to two decimal places.

$$c = \frac{n \times 1000}{V} = \frac{0.00448 \times 1000}{25.00} = 0.18 \text{ mol dm}^{-3}$$

PRACTICE QUESTIONS

3 Joanne carries out a titration to find the concentration of some sulfuric acid. He finds that 28.00 cm³ of sulfuric acid was required to neutralise 25.00 cm³ of 0.02 mol dm⁻³ aqueous potassium hydroxide solution.

$$H_2SO_4(aq) + 2KOH(aq) \rightarrow K_2SO_4(aq) + 2H_2O(l)$$

   a Calculate the number of moles of potassium hydroxide used.
   b Calculate the number of moles of sulfuric acid used.
   c Calculate the concentration of the sulfuric acid. Give your answer to four decimal places.

4 Kevin carries out a titration to find the concentration of some nitric acid. She finds that 28.00 cm³ of nitric acid was required to neutralise 50.00 cm³ of 2.00 mol dm⁻³ aqueous potassium hydroxide solution.

$$HNO_3(aq) + KOH(aq) \rightarrow KNO_3(aq) + H_2O(l)$$

   a Calculate the number of moles of potassium hydroxide used.
   b Calculate the number of moles of nitric acid used.
   c Calculate the concentration of the nitric acid. Give your answer to four decimal places.

STRETCH YOURSELF!

Finding the volume of solution used

In titrations, chemists must find out exactly how much solution is required to neutralise a measured volume of a second solution. This means that they must repeat the titration until they are confident that they have found the correct volume. Chemists know they have found that volume when they repeat the titration and get two very similar (concordant) results. The mean of the concordant results is used to find the volume that is then used in calculations.

Use the results table below to work out the mean volume of sulfuric acid used.

Experiment	Volume of sulfuric acid used / cm³
1	22.80
2	22.50
3	22.60

The mean should only include the concordant results $= \frac{(22.50 + 22.60)}{2} = 22.55 \text{ cm}^3$

PRACTICE QUESTION

5 Use the results table below to work out the mean volume of sodium hydroxide used.

Experiment	Volume of sodium hydroxide used / cm³
1	26.90
2	26.40
3	26.30

1.10 Mole calculations 1

Reacting masses and gas volumes

The balanced equation for a reaction shows how many moles of each reactant and product are involved in a chemical reaction.

Sodium reacts with chlorine to form sodium chloride:

$$2Na(s) + Cl_2(g) \rightarrow 2NaCl(s)$$

The balanced equation shows that 2 moles of sodium react with 1 mole of chlorine molecules to form 2 moles of sodium chloride.

The molar reacting quantities can be calculated using the balanced equation.

If the amount, in moles, of one of the reactants or products is known, the number of moles of any other reactants or products can be calculated.

The number of moles (n), the mass of the substance (m), and the molar mass (M) are linked together using $n = \dfrac{m}{M}$.

The amount of substance (n) in moles and the volume of gas (V) in dm³ are linked together by the equation:

$$n = \frac{V \text{ (in dm}^3)}{24.0}$$

✓ WORKED EXAMPLE

Using balanced equations

Magnesium chloride

Leo reacted 0.243 g of magnesium with chlorine to produce magnesium chloride.

$$Mg(s) + Cl_2(g) \rightarrow MgCl_2(s)$$

Molar mass of $MgCl_2$ is 95.3 g mol⁻¹.

a Calculate the amount, in moles, of magnesium that reacts.

$$n = \frac{m}{M} = \frac{0.243}{24.3} = 0.01 \text{ mol}$$

b Calculate the amount, in moles, of magnesium chloride that was made.

From the balanced equation, the number of moles of magnesium = number of moles of magnesium chloride = 0.01 mol

c Calculate the mass, in grams, of magnesium chloride made. Give your answer to three decimal places.

$$m = n \times M = 0.01 \times 95.3 = 0.953 \text{ g}$$

? PRACTICE QUESTIONS

1 In a reaction, 0.486 g of magnesium was added to oxygen to produce magnesium oxide.

$$2Mg(s) + O_2(g) \rightarrow 2MgO(s)$$

a Calculate the amount, in moles, of magnesium that reacted.

b Calculate the amount, in moles, of magnesium oxide made.

c Calculate the mass, in grams, of magnesium oxide made.

2 In a reaction, 0.115 g of sodium was added to chlorine to produce sodium chloride.

$$2Na(s) + Cl_2(g) \rightarrow 2NaCl(s)$$

a Calculate the amount, in moles, of sodium that reacted.

b Calculate the amount, in moles, of sodium chloride made.

c Calculate the mass, in grams, of sodium chloride made.

3 In a reaction, 0.1955 g of potassium was added to chlorine to produce potassium chloride.

$$2K(s) + Cl_2(g) \rightarrow 2KCl(s)$$

  a Calculate the amount, in moles, of potassium that reacted.

  b Calculate the amount, in moles, of potassium chloride made.

  c Calculate the mass, in grams, of potassium chloride made.

WORKED EXAMPLE

Ciara heated 2.50 g of calcium carbonate, which decomposed as shown in the equation:

$$CaCO_3(s) \rightarrow CaO(s) + CO_2(g)$$

a Calculate the amount, in moles, of calcium carbonate that decomposes.

$$n = \frac{m}{M} = \frac{2.50}{100.1} = 0.025\,mol$$

b Calculate the amount, in moles, of carbon dioxide that forms.

From the balanced equation, the number of moles of calcium carbonate = number of moles of carbon dioxide = 0.025 mol

c Calculate the volume, in dm³, of carbon dioxide made.

V (in dm³) $= n \times 24.0 = 0.025 \times 24 = 0.60\,dm^3$

PRACTICE QUESTIONS

4 Oscar heated 4.25 g of sodium nitrate. The equation for the decomposition of sodium nitrate is given below.

$$2NaNO_3(s) \rightarrow 2NaNO_2(s) + O_2(g)$$

  a Calculate the amount, in moles, of sodium nitrate that reacted.

  b Calculate the amount, in moles, of oxygen made.

  c Calculate the volume, in dm³, of oxygen made at RTP.

5 A 0.2764 g sample of potassium carbonate thermally decomposes to form potassium oxide and carbon dioxide.

$$K_2CO_3(s) \rightarrow K_2O(s) + CO_2(g)$$

  a Calculate the amount, in moles, of potassium carbonate that reacted.

  b Calculate the amount, in moles, of carbon dioxide made.

  c Calculate the volume, in dm³, of carbon dioxide made at RTP.

6 A chemist heated 2.022 g of potassium nitrate. The equation for the decomposition of potassium nitrate is:

$$2KNO_3(s) \rightarrow 2KNO_2(s) + O_2(g)$$

  a Calculate the amount, in moles, of potassium nitrate that reacted.

  b Calculate the amount, in moles, of oxygen made.

  c Calculate the volume, in dm³, of oxygen made at RTP.

7 0.500 kg of magnesium carbonate decomposes on heating to form magnesium oxide and carbon dioxide. Give your answers to three significant figures.

$$MgCO_3(s) \rightarrow MgO(s) + CO_2(g)$$

  a Calculate the amount, in moles, of magnesium carbonate used.

  b Calculate the amount, in moles, of carbon dioxide produced.

  c Calculate the volume, in dm³, of carbon dioxide produced at RTP.

1.11 Mole calculations 2

Reacting masses and volumes

The balanced symbol equation for a reaction can be used to work out the quantities of reactant and products involved in a reaction.

The number of moles (n), the mass of the substance (m), and the molar mass (M) are linked together using $n = \dfrac{m}{M}$.

The amount of substance (n) in moles, the volume of solution (V) in cm^3, and the concentration (c) are linked together by the equation:

$$n = \frac{V \text{ (in } cm^3)}{1000} \times c$$

In these calculations, first work out the amount in moles of one of the substances involved in the reaction. Then use the balanced symbol equation to work out the amount in moles of the desired substances. Then use this information to solve the last part of the question.

WORKED EXAMPLE

Calculating the volume of solutions

Calcium carbonate

Calcium carbonate reacts with $0.25\,mol\,dm^{-3}$ hydrochloric acid to make calcium chloride.

Water and carbon dioxide are also produced. $1.25\,g$ of calcium carbonate is used in the reaction.

$$CaCO_3(s) + 2HCl(aq) \rightarrow CaCl_2(aq) + H_2O(l) + CO_2(g)$$

a Calculate the amount, in moles, of calcium carbonate that reacts.

$$n = \frac{m}{M} = \frac{1.25}{100.1} = 0.0125\,mol$$

b Calculate the amount, in moles, of hydrochloric acid that reacts.

From the balanced equation the number of moles of hydrochloric acid = number of moles of calcium carbonate $\times 2 = 0.0250\,mol$.

c Calculate the volume, in cm^3, of hydrochloric acid used. Give your answer to three significant figures.

$$V = \frac{n \times 1000}{c} = \frac{0.0250 \times 1000}{0.25} = 100\,cm^3$$

PRACTICE QUESTIONS

1 Gordon reacted $2.47\,g$ of copper carbonate with $0.1\,mol\,dm^{-3}$ sulfuric acid.

$$CuCO_3(s) + H_2SO_4(aq) \rightarrow CuSO_4(aq) + H_2O(l) + CO_2(g)$$

   a Calculate the amount, in moles, of copper carbonate that reacted.

   b Calculate the amount, in moles, of sulfuric acid used in the reaction.

   c Calculate the volume, in cm^3, of sulfuric acid used.

2 In a reaction, $1.68\,g$ of magnesium carbonate was reacted with $2.0\,mol\,dm^{-3}$ hydrochloric acid.

$$MgCO_3(s) + 2HCl(aq) \rightarrow MgCl_2(aq) + H_2O(l) + CO_2(g)$$

   a Calculate the amount, in moles, of magnesium carbonate that reacted.

   b Calculate the amount, in moles, of hydrochloric acid used in the reaction.

   c Calculate the volume, in cm^3, of hydrochloric acid used.

3 In a reaction, $1.254\,g$ of zinc carbonate was reacted with $1.0\,mol\,dm^{-3}$ hydrochloric acid.

$$ZnCO_3(s) + 2HCl(aq) \rightarrow ZnCl_2(aq) + H_2O(l) + CO_2(g)$$

   a Calculate the amount, in moles, of zinc carbonate that reacted.

   b Calculate the amount, in moles, of hydrochloric acid used in the reaction.

   c Calculate the volume, in cm^3, of hydrochloric acid used.

WORKED EXAMPLE

Calculating the concentration of solutions

Reacting masses and concentrations

Katie reacted 0.403 g of magnesium oxide with 25.0 cm³ of nitric acid to form magnesium nitrate and water.

$$MgO(s) + 2HNO_3(aq) \rightarrow Mg(NO_3)_2(aq) + H_2O(l)$$

a Calculate the amount, in moles, of magnesium oxide that reacted.

$$n = \frac{m}{M} = \frac{0.403}{40.3} = 0.01 \, mol$$

b Calculate the amount, in moles, of nitric acid used.

From the balanced equation, the amount of nitric acid = 0.02 mol

c Calculate the concentration, in mol dm⁻³, of the nitric acid used.

$$c = \frac{n \times 1000}{v} = \frac{0.02 \times 1000}{25.0} = 0.08 \, mol \, dm^{-3}$$

PRACTICE QUESTIONS

4 Jen reacted 4.00 g of calcium carbonate with exactly 50 cm³ of hydrochloric acid. Calcium chloride, water, and carbon dioxide were produced in the reaction.

$$CaCO_3(s) + 2HCl(aq) \rightarrow CaCl_2(aq) + H_2O(l) + CO_2(g)$$

    a Calculate the amount, in moles, of calcium carbonate that reacted.

    b Calculate the amount, in moles, of hydrochloric acid that reacted.

    c Calculate the concentration of hydrochloric acid used.

5 Tristan reacted 1.686 g of magnesium carbonate with 50 cm³ of hydrochloric acid. Magnesium chloride, water, and carbon dioxide were produced in the reaction.

$$MgCO_3(s) + 2HCl(aq) \rightarrow MgCl_2(aq) + H_2O(l) + CO_2(g)$$

    a Calculate the amount, in moles, of magnesium carbonate that reacted.

    b Calculate the amount, in moles, of hydrochloric acid that reacted.

    c Calculate the concentration of the hydrochloric acid used.

6 Priti reacted 5.00 g of calcium carbonate with 50 cm³ of sulfuric acid. Calcium sulfate, water, and carbon dioxide were produced in the reaction.

$$CaCO_3(s) + H_2SO_4(aq) \rightarrow CaSO_4(aq) + H_2O(l) + CO_2(g)$$

    a Calculate the amount, in moles, of calcium carbonate that reacted.

    b Calculate the amount, in moles, of sulfuric acid that reacted.

    c Calculate the concentration of the sulfuric acid used.

7 A chemist prepared an aqueous solution of 0.10 mol dm⁻³ potassium hydroxide. What is the concentration of the solution in g dm⁻³?

8 A lab technician prepared an aqueous solution of 0.20 mol dm⁻³ sodium hydroxide. What is the concentration of the solution in g dm⁻³?

9 A chemistry teacher prepared an aqueous solution of 0.05 mol dm⁻³ potassium hydroxide. What is the concentration of the solution in g dm⁻³? Give your answer to three significant figures.

1.12 Empirical formula 1

The empirical formula

The empirical formula of a compound is the simplest whole number ratio of atoms of each of the elements present in the compound.

It can be calculated from the mass or the percentage composition by mass of each of the elements present in the compound. The amount of substance (n) in moles, the mass of the substance (m), and the molar mass (M) are linked together using $n = \dfrac{m}{M}$.

WORKED EXAMPLE

Empirical formula from given masses

A sample of titanium oxide was analysed and found to contain 0.958 g of titanium and 0.640 g of oxygen. Determine the empirical formula of the compound. (A_r: Ti, 47.9; O, 16.0)

	Ti	O
mass of each element	0.958	0.640
molar ratio of elements	$\dfrac{0.958}{47.9} = 0.02$	$\dfrac{0.640}{16} = 0.04$
next divide by the smallest number	$\dfrac{0.02}{0.02} = 1$	$\dfrac{0.04}{0.02} = 2$

The empirical formula is TiO_2.

PRACTICE QUESTIONS

1 A sample of phosphorus chloride was analysed and found to contain 0.62 g of phosphorus and 3.55 g of chlorine. Determine the empirical formula of the compound.
(A_r: P, 31.0; Cl, 35.5)

2 A sample of a compound was analysed and found to contain 0.070 g of nitrogen and 0.015 g of hydrogen. Determine the empirical formula of the compound.
(A_r: N, 14.0; H, 1.0)

WORKED EXAMPLE

Analysing a salt

A sample of a salt was analysed and found to contain 0.115 g of sodium, 0.070 g of nitrogen, and 0.240 g of oxygen. Determine the empirical formula of the compound.
(A_r: Na, 23.0; N, 14.0; O, 16.0)

	Na	N	O
mass of each element	0.115	0.070	0.240
molar ratio of elements	$\dfrac{0.115}{23.0} = 0.005$	$\dfrac{0.070}{14.0} = 0.005$	$\dfrac{0.240}{16} = 0.015$
next divide by the smallest number	$\dfrac{0.005}{0.005} = 1$	$\dfrac{0.005}{0.005} = 1$	$\dfrac{0.015}{0.005} = 3$

The empirical formula is $NaNO_3$.

PRACTICE QUESTIONS

3 A sample of a metal carbonate was analysed and found to contain 0.162 g of magnesium, 0.080 g of carbon, and 0.320 g of oxygen. Determine the empirical formula of the compound.
(A_r: Mg, 24.3; C, 12.0; O, 16.0)

4 A sample of a compound was analysed and found to contain 0.254 g of copper, 0.128 g of sulfur, and 0.256 g of oxygen. Determine the empirical formula of the compound.
(A_r: Cu, 63.5; S, 32.1; O, 16.0)

Percentage composition by mass and empirical formulae

Sometimes the analysis of a compound is given as a percentage composition by mass. The empirical formula is calculated using the same method. Simply use the percentage mass by composition instead of the mass in the calculations.

WORKED EXAMPLE

Analysing a compound

A chemist analysed a sample of a compound to find the percentage composition by mass of each element. They found that it contained 20.2% magnesium, 26.7% sulfur, and 53.1% oxygen.

Determine the empirical formula of the compound.
(A_r: Mg, 24.3; S, 32.1; O, 16.0)

	Mg	S	O
percentage composition by mass	20.2	26.7	53.1
molar ratio of elements	$\frac{20.2}{24.3} = 0.831$	$\frac{26.7}{32.1} = 0.832$	$\frac{53.1}{16.0} = 3.319$
next divide by the smallest number	$\frac{0.831}{0.831} = 1$	$\frac{0.832}{0.831} \approx 1$	$\frac{3.319}{0.831} \approx 4$

The empirical formula is $MgSO_4$.

STRETCH YOURSELF!

Information in names

The name of a compound reveals the elements it is made from. For example, a metal oxide will contain metal and oxygen only. If the mass of the sample is known and the mass of the metal is known:

mass of oxygen = mass of compound − mass of metal

This mass can then be used in calculations.

A 0.286 g sample of a metal oxide was analysed and found to contain 0.254 g of copper. Calculate the empirical formula of the compound. (A_r: Cu, 63.5; O, 16.0)

The mass of oxygen in this sample = 0.286 − 0.254 = 0.032 g

	Cu	O
mass of each element	0.254	0.032
molar ratio of elements	$\frac{0.254}{63.5} = 0.004$	$\frac{0.032}{16.0} = 0.002$
next divide by the smallest number	$\frac{0.004}{0.002} = 2$	$\frac{0.002}{0.002} = 1$

The empirical formula is Cu_2O.

PRACTICE QUESTIONS

5 A 0.798 g sample of a metal oxide was analysed and found to contain 0.558 g of iron. Determine the empirical formula of the compound. (A_r: Fe, 55.8; O, 16.0)

6 A 1.268 g sample of a metal chloride was analysed and found to contain 0.558 g of iron. Determine the empirical formula of the compound. (A_r: Fe, 55.8; Cl, 35.5)

1.13 Empirical formula 2

Empirical formula and molecular formula

The empirical formula of a compound is the simplest whole number ratio of atoms of each of the elements present in the compound.

The molecular formula is the actual number of atoms of each element in one molecule.

The molecular formula can be calculated from the empirical formula and the molar mass. If the empirical formula in grams is equal to the molar mass, then the molecular formula and empirical formula are the same.

Empirical formulae can also be calculated by analysing the products of the combustion of compounds. Complete combustion of carbon produces carbon dioxide, whilst combustion of hydrogen produces water vapour.

WORKED EXAMPLES

1 A sample was analysed and found to contain 0.048 g of carbon and 0.016 g of hydrogen.

a Determine the empirical formula of the compound. (A_r: H, 1.0; C, 12.0)

	C	H
mass of each element	0.048	0.016
molar ratio of elements	$\dfrac{0.048}{12.0}$	$\dfrac{0.016}{1.0}$
	$= 0.004$	$= 0.016$
next divide by the smallest number	$\dfrac{0.004}{0.004}$	$\dfrac{0.016}{0.004}$
	$= 1$	$= 4$

The empirical formula is CH_4.

b The compound has a molar mass of 16.0 g mol^{-1}. Find the molecular formula of the compound.

The empirical formula: mass = molecular mass

So the molecular formula is the same as the empirical formula, CH_4.

2 A second sample was analysed and found to contain 0.240 g of carbon and 0.040 g of hydrogen.

a Determine the empirical formula of the compound. (A_r: H, 1.0; C, 12.0)

	C	H
mass of each element	0.24	0.040
molar ratio of elements	$\dfrac{0.024}{12.0}$	$\dfrac{0.040}{1.0}$
	$= 0.020$	$= 0.040$
next divide by the smallest number	$\dfrac{0.020}{0.020}$	$\dfrac{0.040}{0.020}$
	$= 1$	$= 2$

The empirical formula is CH_2.

b The compound has a molar mass of 28.0 g mol^{-1}. Find the molecular formula of the compound.

The empirical formula: mass × 2 = molecular mass

So, the molecular formula is C_2H_4.

PRACTICE QUESTIONS

1 A sample was analysed and found to contain 0.300 g of carbon and 0.050 g of hydrogen.

    a Determine the empirical formula of the compound. (A_r: H, 1.0; C, 12.0)

    b The compound has a molar mass of 84.0 g mol^{-1}. Find the molecular formula of the compound.

2 A sample was analysed and found to contain 0.28 g of nitrogen and 0.04 g of hydrogen.

    a Determine the empirical formula of the compound. (A_r: H, 1.0; N, 14.0)

    b The compound has a molar mass of 32.0 g mol^{-1}. Find the molecular formula of the compound.

3 A sample was analysed and found to contain 0.18 g of carbon and 0.03 g of hydrogen.

    a Determine the empirical formula of the compound. (A_r: H, 1.0; C, 12.0)

    b The compound has a molar mass of 98.0 g mol^{-1}. Find the molecular formula of the compound.

4 A sample was analysed and found to contain 0.16 g of oxygen and 0.01 g of hydrogen.

    a Determine the empirical formula of the compound. (A_r: H, 1.0; O, 16.0)

    b The compound has a molar mass of 34.0 g mol^{-1}. Find the molecular formula of the compound.

STRETCH YOURSELF!

Combustion analysis

Organic compound A was analysed and found to contain hydrogen and carbon only.

A sample of compound A was burnt in excess oxygen and the products of combustion were analysed and found to contain 0.220 g of carbon dioxide and 0.180 g of water vapour.

Determine the empirical formula of compound A.

The molar mass of carbon dioxide = 44.0 g mol^{-1}

The number of moles (n), the mass of the substance (m), and the molar mass (M) are linked together using $n = \dfrac{m}{M}$.

The amount of carbon dioxide, in moles $= \dfrac{0.220\,g}{44.0\,g\,mol^{-1}} = 0.005\,mol$

The molar mass of water = 18.0 g mol^{-1}

The amount of water, in moles $= \dfrac{0.180\,g}{18.0\,g\,mol^{-1}} = 0.010\,mol$

Each molecule of carbon dioxide, CO_2, contains one atom of carbon so compound A must contain 0.005 moles of carbon.

Each molecule of water vapour, H_2O, contains two atom of hydrogen so compound A must contain $0.010 \times 2 = 0.020$ moles of hydrogen.

	C	H
molar ratio of elements	0.005	0.020
next divide by the smallest number	$\dfrac{0.005}{0.005}$	$\dfrac{0.020}{0.005}$
	= 1	= 4

The empirical formula of compound A is CH_4.

PRACTICE QUESTION

5 A compound X was analysed and found to contain hydrogen and carbon only. A sample of compound X was burnt in excess oxygen and the products of combustion were analysed and found to contain 0.176 g of carbon dioxide and 0.054 g of water vapour. Determine the empirical formula of compound X.

1.14 Water of crystallisation

Hydrated salts

Hydrated salts are crystalline compounds that contain water molecules, for example, hydrated copper(II) sulfate. Hydrated copper(II) sulfate has the chemical formula $CuSO_4.5H_2O$. The last part of the name refers to the water of crystallisation. This is the water molecules that are found within the crystalline structure of the hydrated salt. Copper(II) sulfate also exists in an anhydrous form. The anhydrous form of the compound does not contain water molecules.

The molar mass of anhydrous copper(II) sulfate $= 159.6\,\text{g mol}^{-1}$

The molar mass of hydrated copper(II) sulfate $= 159.6 + (5 \times 18.0) = 249.6\,\text{g mol}^{-1}$

Determining the chemical formula of a hydrated salt

When a sample of a hydrated salt is heated strongly, the water of crystallisation can be driven off and evaporated. As this happens the mass of the sample decreases. The sample can be repeatedly heated and then its mass measured until, eventually, the mass remains constant. When this happens all of the water of crystallisation has been removed from the hydrated salt.

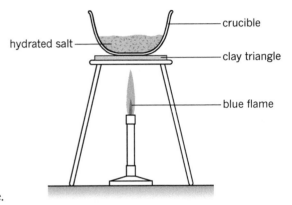

This method is only suitable for salts where the anhydrous form of the compound does not decompose on heating.

The table below shows a sample set of readings from an experiment to find the formula of hydrated sodium carbonate.

Mass of sample before heating / g	5.720
Mass of sample after heating / g	2.120
Mass of water lost during heating / g	3.600

Notice how all the readings are given to the same number of decimal places.

The mass of the sample before heating is the mass of the hydrated salt.

The mass of the sample after heating is the mass of the anhydrous salt.

 WORKED EXAMPLES

Finding the formula of a hydrated salt

Use the information in the table above to find the formula of the hydrated salt.
Number of moles (n), the mass of the substance (m), and the molar mass (M) are linked together using $n = \dfrac{m}{M}$.

The amount, in moles, of anhydrous sodium carbonate $= \dfrac{2.120}{106} = 0.02\,\text{mol}$

The amount, in moles, of water $= \dfrac{3.600}{18.0} = 0.20\,\text{mol}$

The molar ratio of $Na_2CO_3 : H_2O$ is:

	Na_2CO_3	H_2O
molar ratio	0.02	0.20
next divide by the smallest number	$\dfrac{0.02}{0.02}$	$\dfrac{0.20}{0.02}$
	$= 1$	$= 10$

The formula of hydrated sodium carbonate is $Na_2CO_3.10H_2O$.

PRACTICE QUESTIONS

1 A sample of hydrated magnesium sulfate was strongly heated. The mass of the magnesium sulfate before and after heating was recorded in the table below.

Mass of sample before heating / g	1.232
Mass of sample after heating / g	0.602
Mass of water lost during heating / g	

   a Copy and complete the table to show the mass of water lost during heating.
   b Calculate the amount, in moles, of anhydrous magnesium sulfate produced.
   c Calculate the amount, in moles, of water.
   d Calculate the formula of hydrated magnesium sulfate.

2 A sample of hydrated zinc sulfate was strongly heated. The mass of the zinc sulfate before and after heating was recorded in the table below.

Mass of sample before heating / g	2.875
Mass of sample after heating / g	1.615
Mass of water lost during heating / g	

   a Copy and complete the table to show the mass of water lost during heating.
   b Calculate the amount, in moles, of anhydrous zinc sulfate produced.
   c Calculate the amount, in moles, of water.
   d Calculate the formula of hydrated zinc sulfate.

3 A sample of hydrated magnesium chloride was strongly heated. The mass of the magnesium chloride before and after heating was recorded in the table below.

Mass of sample before heating / g	3.706
Mass of sample after heating / g	1.906
Mass of water lost during heating / g	

   a Copy and complete the table to show the mass of water lost during heating.
   b Calculate the amount, in moles, of anhydrous magnesium sulfate produced.
   c Calculate the amount, in moles, of water.
   d Calculate the formula of hydrated magnesium chloride.

STRETCH YOURSELF!

Empirical formula and dot formula

As well as being found in waters of crystallisation, oxygen is also found in sulfate, SO_4^{2-}, carbonate, CO_3^{2-}, and nitrate, NO_3^-, ions.
Chemists use the hydrogen atoms to work out the number of waters of crystallisation.

Worked example

A compound of a hydrated salt has an empirical formula of $CuSH_{10}O_9$.
Write the dot formula for the compound.
If there are 10 hydrogen atoms, there are 5 waters of crystallisation ($5H_2O$).
This leaves $CuSO_4$ so the dot formula is $CuSO_4.5H_2O$.

PRACTICE QUESTION

4 Write the dot formula for a hydrated salt with the empirical formula:
   a $CoCl_2H_{12}O_6$
   b $CaN_2H_8O_{10}$
   c $Na_2SH_{20}O_{14}$.

1.15 Unstructured calculations

Calculations involving a series of steps

Unstructured calculations involve working through a series of steps to get to the final answer.

Read all the information given in the question and identify what you need to work out. Next identify all the information given in the question. Then decide on the best strategy to get to the answer you require.

Molar mass

The molar mass is the mass per mole of a substance. It has the unit of $g\,mol^{-1}$.

WORKED EXAMPLE

Calculating the molar mass of a gas using the ideal gas equation

Example 1

At a temperature of 65°C and a pressure of 125 kPa, a 0.304 g sample of a gas has a volume of 60.0 cm³.

Calculate the molar mass of the gas. Give your answer to three significant figures.

$$P = 125\,kPa = 125 \times 10^3\,Pa$$
$$V = 60.0\,cm^3 = 60 \times 10^{-6}\,m^3$$
$$T = 65°C = 65 + 273 = 338\,K$$
$$R = 8.31$$
$$n = \frac{PV}{RT} = \frac{125 \times 10^3 \times 60 \times 10^{-6}}{8.31 \times 338}$$
$$n = 0.002\,670\,198$$
$$M = \frac{m}{n} = \frac{0.304}{0.002\,670\,198}$$
$$= 114\,g\,mol^{-1}$$

Example 2

A 0.120 g sample of a gas has a volume of 55.0 cm³ and a pressure of 110 kPa at 80°C.

Calculate the molar mass of the gas. Give your answer to two significant figures.

$$P = 110\,kPa = 110 \times 10^3\,Pa$$
$$V = 55.0\,cm^3 = 55 \times 10^{-6}\,m^3$$
$$T = 80°C = 80 + 273 = 353\,K$$
$$R = 8.31$$
$$n = \frac{PV}{RT} = \frac{110 \times 10^3 \times 55 \times 10^{-6}}{8.31 \times 353}$$
$$n = 0.002\,062\,432$$
$$M = \frac{m}{n} = \frac{0.120}{0.002\,062\,432}$$
$$= 58\,g\,mol^{-1}$$

PRACTICE QUESTIONS

1   A 0.256 g sample of a gas has a volume of 75.0 cm³ and a pressure of 115 kPa at 75 °C.

    Calculate the molar mass of the gas. Give your answer to two significant figures.

    $R = 8.31$

2   At a temperature of 50 °C and a pressure of 145 kPa, a 0.316 g sample of a gas has a volume of 75.0 cm³

    Calculate the molar mass of the gas. Give your answer to two significant figures.

    $R = 8.31$

WORKED EXAMPLE

Calculating the concentration of solutions

Amar carries out a titration using the reaction:

$KOH + HCl \rightarrow KCl + H_2O$

He finds 25.0 cm³ of potassium hydroxide reacts with 21.20 cm³ of hydrochloric acid.
The concentration of the hydrochloric acid is 0.005 mol dm⁻³.
What is the concentration of the potassium hydroxide?
Give your answer to three significant figures.

Amount of HCl $= \dfrac{21.20 \times 0.005}{1000} = 0.000\,106$ mol

Amount of KOH $= 0.000\,106$ mol

Concentration of KOH $= \dfrac{0.000\,106 \times 1000}{25.00} = 0.00424$ mol dm⁻³

PRACTICE QUESTIONS

3   Mike carries out a titration to find the concentration of a sample of nitric acid.

    $Na_2CO_3 + 2HNO_3 \rightarrow 2NaNO_3 + H_2O + CO_2$

    The solution of Na_2CO_3 has a concentration of 0.110 mol dm⁻³.

    The volume of Na_2CO_3 is 25.00 cm³.

    The volume of nitric acid required is 23.50 cm³.

    Calculate the concentration of the nitric acid.
    Give your answer to three significant figures.

4   Poppy carries out a titration.

    $Ca(OH)_2 + 2HCl \rightarrow CaCl_2 + 2H_2O.$

    She finds 25.0 cm³ of hydrochloric acid reacts with 28.4 cm³ of calcium hydroxide.

    The concentration of the hydrochloric acid is 0.001 mol dm⁻³.

    What is the concentration of the calcium hydroxide?

    Give your answer to three significant figures.

2.1 Shape of simple molecules

Molecules with covalent bonds

Simple molecules consist of atoms held together by covalent bonds. The shape of these molecules can be determined using electron-pair repulsion theory. Firstly, the central atom is identified. Next, the number of electron pairs surrounding the central atom is counted. These electron pairs repel each other, so the electron pairs are arranged as far away from each other as possible to minimise the repulsion between them. This repulsion determines the shape of the molecule and the angle between the bonds, called the bond angle.

> **REMEMBER:** You will need to learn the names of the mathematical shapes used to describe the molecules.

WORKED EXAMPLE

Molecular shapes and bond angles

Beryllium chloride, BeCl$_2$

The central beryllium atom is surrounded by two pairs of bonded electrons. These electron pairs repel each other so this molecule adopts a linear shape with a bond angle of 180°. Carbon dioxide, CO$_2$, is also linear.

$$Cl \text{——} Be \text{——} Cl \quad 180°$$

Boron trichloride, BCl$_3$

The central boron atom is surrounded by three pairs of bonded electrons. These electron pairs repel each other so this molecule adopts a *trigonal planar* shape with a bond angle of 120°.

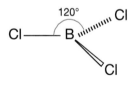

Methane, CH$_4$

The central carbon atom is surrounded by four pairs of bonded electrons. These electron pairs repel each other so this molecule adopts a *tetrahedral* shape with a bond angle of 109.5°. Ammonium, NH$_4^+$, is also tetrahedral.

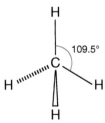

Phosphorus pentachloride, PCl$_5$

The central phosphorus atom is surrounded by five pairs of bonded electrons. These electron pairs repel each other so this molecule adopts a *trigonal bipyramidal* shape with bond angles of 90° and 120°.

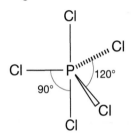

Sulfur hexafluoride, SF$_6$

The central sulfur atom is surrounded by six pairs of bonded electrons. These electron pairs repel each other so this molecule adopts an *octahedral* shape with bond angles of 90°.

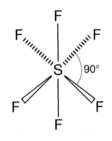

Molecules with lone pairs of electrons

Many molecules contain non-bonding or lone pairs of electrons. Lone pairs of electrons are more electron dense than bonded pairs of electrons so lone pairs repel more than bonded pairs of electrons. This reduces the bond angle in molecules. Each lone pair reduces the bond angle by about 2.5°.

WORKED EXAMPLE

Ammonia, NH₃

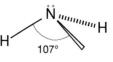

The central nitrogen atom is surrounded by three bonded pairs of electrons and one lone pairs of electrons. The lone pairs repel more than the bonded pairs so the bond angle is reduced to 107°. This shape is called a *trigonal-based pyramidal*.

WORKED EXAMPLE

Water, H₂O

The central oxygen atom is surrounded by two bonded pairs of electrons and two lone pairs of electrons. The lone pairs repel more than the bonded pairs so the bond angle is reduced to 104.5°. This shape is called *non-linear* or *bent*.

PRACTICE QUESTIONS

1 Phosphorus is in the same group as nitrogen. Predict the shape of a phosphine, PH_3, molecule. State its bond angle.

2 Silicon is in the same group as carbon. Name the shape of a silane, SiH_4, molecule. Draw the molecule and label the bond angle.

3 Predict the shape of a phosphorus pentafluoride, PF_5, molecule. State its bond angle.

4 Name the shape of a hydrogen sulfide, H_2S, molecule. Draw the molecule and label the bond angle.

5 Name the shape of an aluminium chloride, $AlCl_3$, molecule. Draw the molecule and label the bond angle.

STRETCH YOURSELF!

Molecules and ions with double bonds

The carbon dioxide, CO_2, molecule contains two double bonds between the carbon atom and the oxygen atoms. Stereochemically, a double bond and a single bond are treated as being equivalent as they repel each other equally. The repulsion between the electron pairs in the two double bonds in the carbon dioxide molecule gives it a linear shape with a bond angle of 180°.

A sulfate, SO_4^{2-}, ion contains two double bonds between the sulfur atom and two of the oxygen atoms, and two single bonds between the sulfur atom and the other two oxygen atoms. The repulsion between the electron pairs in the two double bonds and the two single bonds gives the sulfate ion a tetrahedral shape with a bond angle of 109.5°.

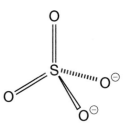

PRACTICE QUESTION

6 Predict the shape of a carbonate, CO_3^{2-}, ion.

3 OXIDATION AND REDUCTION

3.1 Oxidation number

Assigning oxidation numbers

The oxidation number (or state) of an atom is the number of electrons that an atom uses to bond to atoms of other elements.

The oxidation number of an atom cannot be measured directly using experimental data. Instead it is worked out using a few simple rules.

Table 1 The oxidation number rules

Rules	Examples	Exceptions
Any uncombined element has an oxidation number of 0.	Cu, O_2, and S are all uncombined elements so each has an oxidation number of 0.	
Combined fluorine has an oxidation number of -1.	The oxidation number of fluorine in NaF is -1.	
The oxidation number of simple ions is the same as the charge on the ion.	The oxidation number of sodium in Na^+ is $+1$. The oxidation number of calcium in Ca^{2+} is $+2$. The oxidation number of chlorine in Cl^- is -1.	
Combined oxygen has an oxidation number of -2.	The oxidation number of oxygen in H_2O is -2.	If oxygen is combined with fluorine, the oxygen has an oxidation number of $+2$. The oxygen in hydrogen peroxide, H_2O_2, has an oxidation number of -1.
Combined hydrogen has an oxidation number of $+1$.	The oxidation number of hydrogen in H_2O is $+1$.	If hydrogen is combined with a Group 1 metal to form a metal hydride the hydrogen has an oxidation number of -1.
In compounds the overall charge must be zero.		

? PRACTICE QUESTIONS

1 Determine the oxidation number of each element in the following species:

    a Na^+ **b** N_2 **c** MgO   **d** NaF   **e** HCl

2 Determine the oxidation number of each element in the following species:

    a SO_2 **b** H_2O **c** CO   **d** H_2S **e** NO_2

✓ WORKED EXAMPLE

Assigning oxidation numbers in compounds

Calculate the oxidation number of carbon in carbon dioxide.

The overall charge on a carbon dioxide, CO_2, molecule is zero. Therefore the sum of the oxidation numbers of the carbon atom and the two oxygen atoms must be zero. The oxidation number of combined oxygen is -2. As there are two oxygen atoms in the formula this gives a total contribution of -4. This means that the oxidation number of carbon must be $+4$.

Oxidation number in polyatomic ions

Polyatomic ions are groups of covalently bonded atoms that have gained or lost electrons to form ions. Polyatomic ions have an overall charge. The sum of the oxidation numbers of the atoms in the polyatomic ion equals the overall charge on the ion.

WORKED EXAMPLE

SO_4^{2-} ion

Determine the oxidation number of the sulfur in the SO_4^{2-} ion.

The oxidation number of combined oxygen is −2. As there are four combined oxygen atoms this gives an overall contribution of −8.

The overall charge on the ion is −2. This means that the oxidation number of the sulfur must be +6.

PRACTICE QUESTION

3 Determine the oxidation number of each of the underlined elements.

    a $\underline{N}H_4^+$ b $\underline{N}O_3^-$ c $\underline{C}O_3^{2-}$ d $O\underline{H}^-$ e $\underline{S}O_3^{2-}$

WORKED EXAMPLE

Potassium dichromate

Find the oxidation number of chromium in potassium dichromate, $K_2Cr_2O_7$.

The overall charge on the compound is zero and the oxidation numbers apply to each atom.

Potassium has an oxidation number of +1. As there are two potassium atoms this is a total contribution of +2.

Combined oxygen has an oxidation number of −2. As there are seven oxygen atoms this gives a total contribution of −14.

As the overall charge on the compound is zero the two chromium atoms must make a total contribution of +12.

This means that each chromium has an oxidation number of +6.

PRACTICE QUESTION

4 Determine the oxidation number of each of the underlined elements.

    a $\underline{Na}_2O$ b $H_2\underline{S}O_4$ c $\underline{Al}_2O_3$ d $\underline{K}_2O$ e $\underline{H}_2SO_4$

STRETCH YOURSELF!

Transition elements

Transition elements form ions with different oxidation numbers. For example, iron can form either Fe^{2+} or Fe^{3+} ions. When naming compounds that contain transition elements, the oxidation number of the transition element is included in the name using Roman numerals. For example:

Fe_2O_3 is iron(III) oxide as the iron has an oxidation number of +3.

Fe_3O_2 is iron(II) oxide as the iron has an oxidation number of +2.

PRACTICE QUESTION

5 State the oxidation number of the transition element in each of these compounds:

    a titanium(IV) chloride   b copper(I) oxide   c nickel(II) sulfate

    d iron(III) chloride   e copper(II) oxide

3.2 Oxidation and reduction

Oxidation numbers

The oxidation number (or state) of an atom is the number of electrons that an atom uses to bond to atoms of other elements. In oxidation a species loses electrons so there is an increase in its oxidation number. In reduction a species gains electrons so there is a decrease in its oxidation number.

Oxidation and reduction must happen together so these reactions are known as redox reactions.

WORKED EXAMPLE

Oxidation and reduction reactions

Sodium reacts with chlorine to form sodium chloride.

$$2Na + Cl_2 \rightarrow 2NaCl$$

During the reaction, sodium atoms lose electrons to form sodium ions.

$$2Na \rightarrow 2Na^+ + 2e^-$$

The oxidation number of sodium atoms increases from 0 to +1 so the sodium is oxidised in the reaction.

At the same time, chlorine atoms gain electrons to form chloride ions.

$$Cl_2 + 2e^- \rightarrow 2Cl^-$$

The oxidation number of the chlorine atoms decreases from 0 to −1 so the chlorine atoms are reduced in the reaction.

PRACTICE QUESTIONS

1 Zinc reacts with hydrochloric acid to form zinc chloride and hydrogen.
$$Zn + 2HCl \rightarrow ZnCl_2 + H_2$$
Use oxidation numbers to show that zinc is oxidised in this reaction.

2 Magnesium reacts with oxygen to form magnesium oxide.
$$2Mg + O_2 \rightarrow 2MgO$$
Use oxidation numbers to find which species has been:
a oxidised   b reduced.

3 Zinc reacts with oxygen to form zinc oxide.
$$2Zn + O_2 \rightarrow 2ZnO$$
Use oxidation numbers to find which species has been:
a oxidised   b reduced.

4 Magnesium reacts with chlorine to form magnesium chloride.
$$Mg + Cl_2 \rightarrow MgCl_2$$
Use oxidation numbers to find which species has been:
a oxidised   b reduced.

Metals and acids

Some metals react with acids to form a salt and hydrogen. During these reactions the metal atoms lose electrons to form metal ions, so the metal atoms are oxidised. Acids react with water to form H^+ ions. During the reaction with metals the hydrogen ions gain electrons, so the hydrogen in the acid is reduced.

WORKED EXAMPLE
Oxidation and reduction

1 Magnesium reacts with hydrochloric acid to form magnesium chloride and hydrogen.

$$Mg + 2HCl \rightarrow MgCl_2 + H_2$$

Use oxidation numbers to identify what has been oxidised and what has been reduced.

The oxidation number of magnesium has increased from 0 to +2 so the magnesium has been oxidised.

The oxidation number of hydrogen has decreased from +1 to 0 so the hydrogen in the acid has been reduced.

2 Zinc reacts with nitric acid. The reaction is represented by the equation:

$$Zn + 2HNO_3 \rightarrow Zn(NO_3)_2 + H_2$$

Use oxidation numbers to identify what has been oxidised and what has been reduced.

The oxidation number of zinc has increased from 0 to +2 so the zinc has been oxidised.

The oxidation number of hydrogen has decreased from +1 to 0 so the hydrogen in the acid has been reduced.

PRACTICE QUESTIONS

5 Use oxidation numbers to identify what has been oxidised and what has been reduced in this reaction:

$$Mg + 2HNO_3 \rightarrow Mg(NO_3)_2 + H_2$$

6 Use oxidation numbers to identify what has been oxidised and what has been reduced in this reaction:

$$Zn + 2HCl \rightarrow ZnCl_2 + H_2$$

Oxidising and reducing agents

An oxidising agent is a reagent that oxidises another species.
A reducing agent is a reagent that reduces another species.

STRETCH YOURSELF!
Oxidation number of oxyanions

Negatively charged ions that contain oxygen are known as oxyanions. Some elements are able to form more than one type of oxyanion. When naming the ion, the oxidation number of the element bonded to oxygen is given using Roman numerals.

Example

Sulfur can form SO_4^{2-} or SO_3^{2-} ions. Name each of these oxyanions.
SO_4^{2-} contains sulfur atoms with an oxidation number of +6 so the ion is sulfate(VI).
SO_3^{2-} contains sulfur atoms with an oxidation number of +4 so the ion is sulfate(IV).

PRACTICE QUESTION

7 Give the formula of a nitrate(V) ion, which has an overall charge of −1.

3.3 Group 2

Good reducing agents

Group 2 elements are found in the s-block of the periodic table. Atoms of Group 2 elements have two electrons in their outer shell. During chemical reactions, Group 2 metals lose electrons to form ions with an oxidation number of +2. This means that Group 2 metals are good reducing agents.

Reactions with oxygen

Group 2 elements react with oxygen to form metal oxides.

The general equation is $2M + O_2 \rightarrow 2MO$, where M represents the Group 2 element.

Half equations can be used to show clearly what has happened to each species.

$$2M \rightarrow 2M^{2+} + 4e^- \qquad O_2 + 4e^- \rightarrow 2O^{2-}$$

Notice how both the species and the charge must balance in these equations.

WORKED EXAMPLES

Calcium and oxygen

Calcium reacts with oxygen to form calcium oxide.

$$2Ca(s) + O_2(g) \rightarrow 2CaO(s)$$

a Write separate half equations for this reaction.

$$2Ca \rightarrow 2Ca^{2+} + 4e^- \qquad O_2 + 4e^- \rightarrow 2O^{2-}$$

b Identify the species that are oxidised and reduced in the reaction.

The oxidation number of calcium increases from 0 to +2 so the calcium is oxidised.

The oxidation number of oxygen decreases from 0 to −2 so the oxygen is reduced.

Magnesium and oxygen

Magnesium reacts with oxygen to form magnesium oxide.

$$2Mg(s) + O_2(g) \rightarrow 2MgO(s)$$

a Use oxidation numbers to identify which species have been oxidised and reduced in this reaction.

The oxidation number of magnesium increases from 0 to +2 so the magnesium is oxidised.

The oxidation number of oxygen decreases from 0 to −2 so the oxygen is reduced.

b Identify the oxidising agent and the reducing agent in this reaction.

The magnesium reduces the oxygen, so the magnesium is the reducing agent.

The oxygen oxidises the magnesium, so the oxygen is the oxidising agent.

PRACTICE QUESTIONS

1 Strontium reacts with oxygen to form strontium oxide.

$$2Sr + O_2 \rightarrow 2SrO$$

   a Use oxidation numbers to identify which species have been oxidised and reduced in this reaction.

   b Identify the oxidising agent and the reducing agent in this reaction.

2 Barium reacts with oxygen to form barium oxide.

$$2Ba + O_2 \rightarrow 2BaO$$

   a Use oxidation numbers to identify which species have been oxidised and reduced in this reaction.

   b Identify the oxidising agent and the reducing agent in this reaction.

Reactions with water

Group 2 elements react with water to form metal hydroxide solutions and hydrogen. Group 2 elements are strong reducing agents and reduce one of the hydrogen atoms in water during this reaction.

WORKED EXAMPLES

Calcium hydroxide

Calcium reacts with water to form calcium hydroxide and hydrogen.
$$Ca(s) + 2H_2O(l) \rightarrow Ca(OH)_2(aq) + H_2(g)$$

a Use oxidation numbers to identify which species are oxidised and reduced in the reaction.
The oxidation number of calcium increases from 0 to +2 so the calcium is oxidised.
The oxidation number of one of the hydrogen atoms in water decreases from +1 to 0 so the hydrogen is reduced.

b Identify the oxidising agent and the reducing agent.
The calcium reduces the hydrogen, so the calcium is the reducing agent.
The water oxidises the calcium, so the water is the oxidising agent.
Notice how the oxidising agent is the name of the whole reagent – water.

PRACTICE QUESTIONS

3 Strontium reacts with water to form strontium hydroxide. Hydrogen is also produced.
$$Sr(s) + 2H_2O(l) \rightarrow Sr(OH)_2(aq) + H_2(g)$$

    a Use oxidation numbers to identify which species are oxidised and reduced in the reaction.

    b Identify the oxidising agent and the reducing agent.

4 Barium reacts with water to form barium hydroxide. Hydrogen is also produced.
$$Ba(s) + 2H_2O(l) \rightarrow Ba(OH)_2(aq) + H_2(g)$$

    a Use oxidation numbers to identify which species are oxidised and reduced in the reaction.

    b Identify the oxidising agent and the reducing agent.

WORKED EXAMPLE

A 1.5 g sample of calcium carbonate decomposes to form calcium oxide and carbon dioxide.
$$CaCO_3(s) \rightarrow CaO(s) + CO_2(g)$$

a Calculate the amount, in moles, of calcium carbonate used.
$$n = \frac{m}{M} = \frac{1.5}{100.1} = 0.015 \, mol$$

b Calculate the amount, in moles, of carbon dioxide made.
Amount, in moles, of calcium oxide = amount, in moles, of carbon dioxide = 0.015 mol

c Calculate the volume at RTP, in cm³, of carbon dioxide made.
$$V = n \times 24\,000 = = 0.015 \times 24\,000 \, cm^3 = 360 \, cm^3$$

PRACTICE QUESTION

5 A 0.738 g sample of strontium carbonate decomposes to form strontium oxide and carbon dioxide.
$$SrCO_3(s) \rightarrow SrO(s) + CO_2(g)$$

    a Calculate the amount, in moles, of stontium carbonate used.

    b Calculate the amount, in moles, of carbon dioxide made.

    c Calculate the volume at RTP, in cm³, of carbon dioxide made.

3.4 Group 7

Good oxidising agents

Group 7 elements are called halogens. They are found in the p-block of the periodic table. Atoms of Group 7 elements have seven electrons in their outer shell. During chemical reactions, atoms of Group 7 elements typically gain one electron to form halide ions with an oxidation number of -1. This means that Group 7 elements are good oxidising agents.

Reactions with metals

Group 7 elements react with metals to form metal halides.

WORKED EXAMPLE

Potassium and chlorine

Potassium reacts with chlorine to form potassium chloride.

$$2K(s) + Cl_2(g) \rightarrow 2KCl(s)$$

a Write separate half equations for this reaction.

$$2K \rightarrow 2K^+ + 2e^-$$

$$Cl_2 + 2e^- \rightarrow 2Cl^-$$

b Use oxidation numbers to identify which species are oxidised and reduced in this reaction.
The oxidation number of potassium increases from 0 to $+1$ so the potassium is oxidised.
The oxidation number of chlorine decreases from 0 to -1 so the chlorine is reduced.

c Identify the oxidising agent and the reducing agent in this reaction.
The potassium reduces the chlorine, so the potassium is the reducing agent.
The chlorine oxidises the potassium, so the chlorine is the oxidising agent.

PRACTICE QUESTIONS

1 A chemist reacts sodium with fluorine to form sodium fluoride.

$$2Na(s) + F_2(g) \rightarrow 2NaF(s)$$

  a Use oxidation numbers to identify which species have been oxidised and reduced in this reaction.

  b Identify the oxidising agent and the reducing agent in this reaction.

2 A chemist reacts potassium with fluorine to form potassium fluoride.

$$2K(s) + F_2(g) \rightarrow 2KF(s)$$

  a Use oxidation numbers to identify which species have been oxidised and reduced in this reaction.

  b Identify the oxidising agent and the reducing agent in this reaction.

Redox reactions

Group 7 elements become less reactive going down the group. A more reactive halogen will oxidise a less reactive halogen and displace it from a solution of its salt.

WORKED EXAMPLE

Chlorine reacts with an aqueous solution of potassium bromide to form potassium chloride and bromine.

$$Cl_2(aq) + 2KBr(aq) \rightarrow 2KCl(aq) + Br_2(aq)$$

a Write separate half equations for this reaction.

$2Br^- \rightarrow Br_2 + 2e^-$

$Cl_2 + 2e^- \rightarrow 2Cl^-$

b Use oxidation numbers to identify which species are oxidised and reduced in this reaction.

The oxidation number of bromine increases from -1 to 0 so the bromine is oxidised.

The oxidation number of chlorine decreases from 0 to -1 so the chlorine is reduced.

c Identify the oxidising agent and reducing agent in this reaction.

The bromine reduces the chlorine, so the potassium bromide is the reducing agent.
The chlorine oxidises the bromine, so the chlorine is the oxidising agent.

> REMEMBER: The reducing agent is the name of the whole reagent, e.g., potassium bromide.

PRACTICE QUESTIONS

3 Bromine reacts with an aqueous solution of potassium iodide to form potassium bromide and iodine.

$$Br_2(aq) + 2KI(aq) \rightarrow 2KBr(aq) + I_2(aq)$$

a Use oxidation numbers to identify which species are oxidised and reduced in the reaction.

b Identify the oxidising agent and the reducing agent.

4 Chlorine reacts with an aqueous solution of potassium iodide to form potassium chloride and iodine.

$$Cl_2(aq) + 2KI(aq) \rightarrow 2KCl(aq) + I_2(aq)$$

a Use oxidation numbers to identify which species are oxidised and reduced in the reaction.

b Identify the oxidising agent and the reducing agent.

STRETCH YOURSELF!

Disproportionation reactions

In disproportionation reactions a species is simultaneously oxidised and reduced.

Example

Chlorine reacts with water to form chloric(I) acid and hydrochloric acid.

$$Cl_2(aq) + H_2O(l) \rightarrow HClO(aq) + HCl(aq)$$

Use oxidation numbers to prove that this is a disproportionation reaction.

The oxidation number of chlorine increases from 0 to $+1$ (in chloric(I) acid), whilst the oxidation number of chlorine decreases from 0 to -1 (in hydrochloric acid). So chlorine is simultaneously oxidised and reduced.

PRACTICE QUESTION

5 Copper(I) oxide reacts to form copper(II) oxide and copper.

$$Cu_2O \rightarrow CuO + Cu$$

Use oxidation numbers to show that this is a disproportionation reaction.

1 In an experiment to produce crystals of hydrated copper(II) sulfate a 0.3975 g sample of copper(II) oxide was added to a beaker containing 75 cm³ of 0.10 mol dm⁻³ sulfuric acid. The reaction is summarised below.

$$CuO(s) + H_2SO_4(aq) \rightarrow CuSO_4(aq) + H_2O(l)$$

(A_r: Cu, 63.5; S, 32.1; O, 16; H, 1)

   a Calculate the amount, in moles, of copper(II) oxide added.

   b Calculate the amount, in moles, of sulfuric acid in the beaker.

   c State and explain which substance is in excess.

2 Lukas analysed a sample of lead and found it contained three isotopes. The percentage abundance of these isotopes is shown below.

Isotope	Percentage abundance
206	24.0
207	22.1
208	53.9

Calculate the relative atomic mass of lead. Give your answer to one decimal place.

3 Chlorine is added to swimming pools and hot tubs to prevent the spread of waterborne diseases. The chlorine reacts with water to produce two acids. The reaction is summed up below.

$$Cl_2(aq) + H_2O(l) \rightarrow HClO(aq) + HCl(aq)$$

Explain why this reaction can be described as an example of disproportionation.

4 Sulfur hexafluoride, SF_6, is a covalent substance that is a gas at room temperature.

   a Predict and explain the shape of a molecule of sulfur hexafluoride.

   b Calculate the number of molecules in a 0.73 g sample of sulfur hexafluoride. Give your answer to three significant figures.

   (A_r: S, 32; F, 19)

   Avogadro constant $= 6.02 \times 10^{23}$

5 Compound A was analysed and found to contain only carbon and hydrogen. A sample of compound A was burnt in an excess of oxygen and the products of combustion were collected and analysed. There was 1.100 g of CO_2 and 0.675 g of H_2O.

The molar mass of $CO_2 = 44.0$ g mol⁻¹

The molar mass of $H_2O = 18.0$ g mol⁻¹

Calculate the empirical formula of compound A.

6 Calculate the oxidation number of sulfur in each of these compounds.

   a H_2S

   b H_2SO_4

   c SO_2

7 Sulfur dioxide reacts with sodium hydroxide to produce sodium sulfate. The reaction is summarised below.

$SO_2(g) + 2NaOH(aq) \rightarrow Na_2SO_3(aq) + H_2O(l)$

a Calculate the volume of sulfur dioxide, in dm^3, measured at room temperature and pressure that would react completely with $50\,cm^3$ of a solution of $1.00\,mol\,dm^{-3}$ sodium hydroxide.

(Molar volume of a gas $= 24.0\,dm^3\,mol^{-1}$ at room temperature and pressure.)

b Calculate the number of molecules of sulfur dioxide that would react with the sodium hydroxide solution. Give your answer to three significant figures.

(Avogadro's constant $= 6.02 \times 10^{23}$)

8 Phosphorus reacts with chlorine to produce different compounds. One compound is analysed and found to contain 22.5% phosphorus by mass. Calculate the empirical formula of this compound.

(A_r: P, 31; Cl, 35.5)

9 Andrea analysed a sample of nickel and found it contained two isotopes: nickel-58 and nickel-60.

The nickel has a relative atomic mass of 58.62. Calculate the percentage abundance of the two isotopes in her sample.

10 In an experiment to make crystals of zinc sulfate, 4.18g of zinc carbonate is added to an excess of sulfuric acid. The following reaction occurs:

$ZnCO_3 + H_2SO_4 \rightarrow ZnSO_4 + H_2O + CO_2$

(A_r: Zn, 65.4; C, 12.0; S, 32.1; O, 16.0; H, 1.0]

a Calculate the amount, in moles, of zinc carbonate added. Give your answer to three decimal places.

b Calculate the mass of zinc sulfate produced in this reaction. Give your answer to three significant figures.

11 A chemist reacts potassium with chlorine to form potassium chloride.

$2K(s) + Cl_2(g) \rightarrow 2KCl(s)$

a Use oxidation numbers to identify which species have been oxidised and which have been reduced in this reaction.

b Identify the oxidising agent and the reducing agent in this reaction.

12 Use the results table below to work out the mean average volume of sodium hydroxide used. Give your answer to 2 decimal places.

Experiment	Volume of sodium hydroxide used / cm³
1	22.60
2	21.90
3	21.50
4	21.80

4 ENTHALPY CHANGES

4.1 Enthalpy changes 1

Enthalpy change, $\triangle H$

Enthalpy, H, is the heat content that is stored in a chemical system. Enthalpy change, $\triangle H$, is the heat change at constant pressure, which can be calculated from experimental data.

In exothermic reactions, heat is lost by the chemical system and gained by the surroundings so a temperature rise is observed. The enthalpy change for an exothermic reaction has a negative sign.

In endothermic reactions, heat is gained by the chemical system from the surroundings so a temperature fall is observed. The enthalpy change for an endothermic reaction has a positive sign.

Heat exchange

The heat gained or lost by the surroundings can be calculated using the relationship:

$Q = mc\triangle T$

Q is the heat exchange measured in joules, J.

m is the mass of the surroundings involved in the heat exchange measured in grams, g.

c is the specific heat capacity, the amount of energy required to raise the temperature of 1 g of the substance by 1 °C. It is measured in $J\,g^{-1}\,K^{-1}$.

$\triangle T$ is the change in temperature:
the final temperature − the initial temperature (measured in °C).

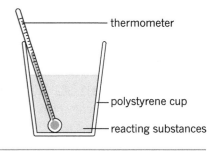

WORKED EXAMPLE

An excess of magnesium is added to 50 cm³ of aqueous copper sulfate solution.
The temperature of the solution increases from 21 °C to 32 °C.

Calculate the heat exchanged in this reaction.

The specific heat capacity of the solution, c, is $4.18\,J\,g^{-1}\,K^{-1}$.

The solution has a density of $1.00\,g\,cm^{-3}$.

$Q = mc\triangle T = 50 \times 4.18 \times 11 = 2299\,J$

PRACTICE QUESTIONS

1 An excess of zinc is added to 100 cm³ of dilute sulfuric acid. The temperature of the solution increases from 19 °C to 24 °C.
Calculate the heat exchanged in this reaction.
Specific heat capacity of the solution, c, is $4.18\,J\,g^{-1}\,K^{-1}$.
The solution has a density of $1.00\,g\,cm^{-3}$.

2 An excess of iron is added to 150 cm³ of dilute copper sulfate solution. The temperature of the solution increases from 18 °C to 32 °C.
Calculate the heat exchanged in this reaction.
Specific heat capacity of the solution, c, is $4.18\,J\,g^{-1}\,K^{-1}$.
The solution has a density of $1.00\,g\,cm^{-3}$.

Determining the enthalpy change of a reaction

The enthalpy change of a reaction can be calculated from the heat exchange and the amount of substance, in moles, that has reacted.

WORKED EXAMPLES

Calculating an enthalpy change

An excess of iron is added to 50 cm^3 of 2.0 mol dm^{-3} aqueous copper(II) sulfate solution.

$$Fe(s) + CuSO_4(aq) \rightarrow FeSO_4(aq) + Cu(s)$$

The temperature of the solution increases from 22 °C to 35 °C.

The specific heat capacity of the solution, c, is 4.18 J g^{-1} K^{-1}.

The solution has a density of 1.00 g cm^{-3}.

a   Find the energy exchanged in this reaction.
$Q = mc\triangle T = 50 \times 4.18 \times 13 = 2717$ J or 2.717 kJ

b   Find the amount, in moles, of copper(II) sulfate solution used in the reaction.
$n = V$ (in dm^3) $\times$ concentration $= 0.05 \times 2.0 = 0.10$ mol

c   Calculate the enthalpy change in kJ mol^{-1}.
To scale up the enthalpy change to kJ mol^{-1} = $\dfrac{\text{energy exchange (kJ)}}{\text{amount of substance (mol)}} = \dfrac{2.717 \text{ kJ}}{0.10 \text{ mol}} = 27.17$ kJ mol^{-1}

As the temperature of the surroundings increased, heat was lost by the chemical system and gained by the surroundings. This means the reaction was exothermic and the enthalpy change of the reaction has a negative sign.

The enthalpy change is -27.17 kJ mol^{-1}.

PRACTICE QUESTIONS

3   An excess of magnesium is added to 100 cm^3 of 1.00 mol dm^{-3} hydrochloric acid.
The temperature increases from 19 °C to 27 °C.
$$Mg(s) + 2HCl(aq) \rightarrow MgCl_2(aq) + H_2(g)$$
The specific heat capacity of the solution, c, is 4.18 J g^{-1} K^{-1}.
The solution has a density of 1.00 g cm^{-3}.
a   Calculate the energy change for the reaction.
b   Calculate the amount, in moles, of hydrochloric acid used.
c   Calculate the enthalpy change for the reaction.

4   An excess of zinc is added to 150 cm^3 of 1.00 mol dm^{-3} nitric acid. The temperature increases from 22 °C to 29 °C.
$$Zn(s) + 2HNO_3(aq) \rightarrow Zn(NO_3)_2(aq) + H_2(g)$$
The specific heat capacity of the solution, c, is 4.18 J g^{-1} K^{-1}.
The solution has a density of 1.00 g cm^{-3}.
a   Calculate the energy change for the reaction.
b   Calculate the amount, in moles, of hydrochloric acid used.
c   Calculate the enthalpy change for the reaction.

STRETCH YOURSELF!

Experimental and theoretical values

The enthalpy change calculated from experimental results is less exothermic or less endothermic than the theoretical value for the reaction. This is due to heat loss from the surroundings. This reduces the temperature change that is detected during the experiment.

PRACTICE QUESTION

5   Suggest how a student could reduce heat loss during an experiment to determine the enthalpy change of the reaction between iron and an aqueous solution of copper(II) sulfate.

4.2 Enthalpy changes 2

Standard enthalpy change of combustion, $\Delta H_c^{\ominus}$

The standard enthalpy change of combustion, $\Delta H_c^{\ominus}$, is the enthalpy change when one mole of a substance is completely burnt in oxygen, under standard conditions, with all reactants and products being in their standard states.

Enthalpy changes of combustion are always exothermic so the sign for enthalpy change of combustion reactions is negative.

The chemical equation for the standard enthalpy change of combustion of methane is:

$$CH_4(g) + 2O_2(g) \rightarrow CO_2(g) + 2H_2O(l)$$

The enthalpy change of combustion can be determined from experimental data.

A known mass of fuel is burnt and heats up a known volume of water. The temperature change of the water is found.

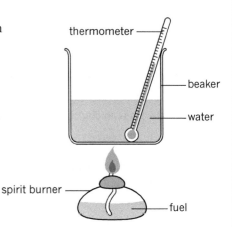

Heat exchange

The heat lost by the system and gained by the surroundings can be calculated using the relationship:

$$Q = mc\Delta T$$

Q is the heat exchange measured in joules, J.

m is the mass of the water involved in the heat exchange measured in grams, g.

c is the specific heat capacity of water, the amount of energy required to raise the temperature of 1 g of the substance by 1 °C. It has a value of $4.18\,J\,g^{-1}\,K^{-1}$.

ΔT is the change in temperature of the water:

ΔT = the final temperature − the initial temperature (measured in °C).

WORKED EXAMPLE

Calculating heat exchange

Burning ethanol

During the combustion of a sample of ethanol, 100 cm³ of water was heated from 21 °C to 35 °C. Calculate the heat exchanged in this reaction.

The specific heat capacity of water, c, is $4.18\,J\,g^{-1}\,K^{-1}$.

Water has a density of $1.00\,g\,mol^{-1}$.

$\Delta T = 35 - 21 = 14\,K$

$Q = mc\Delta T = 100 \times 4.18 \times 14 = 5852\,J$ or $5.852\,kJ$

PRACTICE QUESTIONS

1. Miguel burns a sample of propan-1-ol. During the experiment, the temperature of 250 cm³ of water increases from 18 °C to 24 °C.
   Calculate the heat exchanged in this reaction.
   Specific heat capacity of water, c, is $4.18\,J\,g^{-1}\,K^{-1}$.
   Water has a density of $1.00\,g\,mol^{-1}$.

2. During combustion of a sample of ethanol, the temperature of 150 cm³ of water increased from 24 °C to 39 °C.
   Calculate the heat exchanged in this reaction.
   Specific heat capacity of water, c, is $4.18\,J\,g^{-1}\,K^{-1}$.
   Water has a density of $1.00\,g\,mol^{-1}$.

3 During combustion of a sample of butan-2-ol, the temperature of 200 cm³ of water increased from 15 °C to 27 °C.
Calculate the heat exchanged in this reaction.
Specific heat capacity of water, c, is $4.18\,\mathrm{J\,g^{-1}\,K^{-1}}$.
Water has a density of $1.00\,\mathrm{g\,mol^{-1}}$.

Determining the enthalpy change of combustion

The enthalpy change of a combustion reaction can be calculated from the heat gained by the surroundings and the amount of substance, in moles, that has been burnt.

WORKED EXAMPLE

Calculating the enthalpy change of combustion

During combustion, 0.46 g of ethanol, C_2H_5OH, heated 100 cm³ of water. The temperature of the water increased from 22 °C to 48 °C.

The specific heat capacity of water, c, is $4.18\,\mathrm{J\,g^{-1}\,K^{-1}}$.

Water has a density of $1.00\,\mathrm{g\,cm^{-3}}$.

a Calculate the energy exchanged in this reaction.
$Q = mc\triangle T = 100 \times 4.18 \times 26 = 10\,868\,\mathrm{J}$ or $10.868\,\mathrm{kJ}$

b Find the amount, in moles, of ethanol that was used in this reaction.
Molar mass of ethanol, $C_2H_5OH = 46.0\,\mathrm{g\,mol^{-1}}$
$n = \dfrac{m}{M} = \dfrac{0.46}{46.0} = 0.01\,\mathrm{mol}$

c Calculate the enthalpy change in $\mathrm{kJ\,mol^{-1}}$.
To scale up the enthalpy change to $\mathrm{kJ\,mol^{-1}}$:

$$\frac{\text{energy exchange (kJ)}}{\text{amount of substance (mol)}} = \frac{10.868\,\mathrm{kJ}}{0.01\,\mathrm{mol}} = 1086.8\,\mathrm{kJ\,mol^{-1}}$$

The enthalpy change is $-1086.8\,\mathrm{kJ\,mol^{-1}}$.

d Suggest the significance of the sign for the enthalpy change.
The negative sign shows the reaction is exothermic.

PRACTICE QUESTIONS

4 During combustion, 1.80 g of propan-1-ol, C_3H_7OH, heated 200 cm³ of water. The temperature of the water increased from 20 °C to 51 °C.
The specific heat capacity of water, c is $4.18\,\mathrm{J\,g^{-1}\,K^{-1}}$.
Water has a density of $1.00\,\mathrm{g\,cm^{-3}}$.
Give your answers to three significant figures.
  a Find the heat energy change in this reaction.
  b Find the amount, in moles, of propan-1-ol used.
  c Find the enthalpy change for this reaction in $\mathrm{kJ\,mol^{-1}}$.
  d Suggest the significance of the sign of the enthalpy change.

5 During combustion, 2.30 g of ethanol, C_2H_5OH, heated 100 cm³ of water. The temperature of the water increased from 20 °C to 32 °C.
Specific heat capacity of water, c, is $4.18\,\mathrm{J\,g^{-1}\,K^{-1}}$.
Water has a density of $1.00\,\mathrm{g\,cm^{-3}}$.
Give your answers to three significant figures.
  a Find the heat energy change in this reaction.
  b Find the amount, in moles, of ethanol used.
  c Find the enthalpy change for this reaction in $\mathrm{kJ\,mol^{-1}}$.

4.3 Hess's law 1

Calculating the enthalpy change for a reaction indirectly

Sometimes it is impossible to measure the enthalpy change of a reaction directly. This could be for one or more of the following reasons.

- More than one reaction takes place. For example, it would be very difficult to measure this enthalpy change directly:

    $$2C(s) + 2H_2(g) \rightarrow C_2H_4(g)$$

    because lots of different hydrocarbons are made at the same time.
- The reaction has a very slow rate of reaction.
- The reaction has a very high activation energy.

In such situations, chemists can calculate the enthalpy change for a reaction indirectly using Hess's law. This states that the enthalpy change for a reaction is independent of the route taken, provided the initial conditions and the final conditions are the same.

Standard enthalpy changes of combustion

Standard enthalpy changes of combustion, $\Delta H_c^\ominus$, values can be used in Hess's law calculations.

The standard enthalpy change of combustion, $\Delta H_c^\ominus$, is the enthalpy change that takes place when one mole of a substance is completely burnt in oxygen, all reactants and products being in their standard states.

Notice how the arrows go from the substances to the combustion products.

A is the enthalpy change for the reaction.

B is the enthalpy change of combustion for the reactants.

C is the enthalpy change of combustion for the products.

The direct route (A) is equal to the indirect route (B − C).

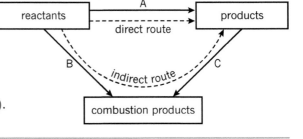

 WORKED EXAMPLE

Determining the enthalpy change for a reaction

Determine the enthalpy change for the reaction between carbon and hydrogen to make methane.

$$C(s) + 2H_2(g) \rightarrow CH_4(g)$$

Use the enthalpy change of combustion values below:

$\Delta H_c^\ominus$ C(s) $= -394\,kJ\,mol^{-1}$
$\Delta H_c^\ominus$ H$_2$(g) $= -286\,kJ\,mol^{-1}$
$\Delta H_c^\ominus$ CH$_4$(g) $= -890\,kJ\,mol^{-1}$

A $= +(-394) + (2 \times -286) - (-890)$
  $= -76\,kJ\,mol^{-1}$

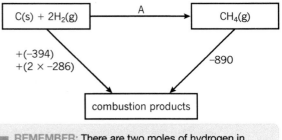

REMEMBER: There are two moles of hydrogen in the equation so the value for the enthalpy change of combustion for hydrogen must be doubled.

? **PRACTICE QUESTIONS**

1   Determine the enthalpy change for the reaction between ethene and hydrogen to make ethane.

    $$C_2H_4(g) + H_2(g) \rightarrow C_2H_6(g)$$

    Use the enthalpy change of combustion values below:

    $\Delta H_c^\ominus$ C$_2$H$_4$(g) $= -1411\,kJ\,mol^{-1}$
    $\Delta H_c^\ominus$ H$_2$(g) $= -286\,kJ\,mol^{-1}$
    $\Delta H_c^\ominus$ C$_2$H$_6$(g) $= -1560\,kJ\,mol^{-1}$

2 Determine the enthalpy change for the reaction between carbon and hydrogen to make butane.

$$4C(s) + 5H_2(g) \rightarrow C_4H_{10}(g)$$

Use the enthalpy change of combustion values below:

$\Delta H_c^\ominus$ C(s) $= -394\,kJ\,mol^{-1}$
$\Delta H_c^\ominus$ H$_2$(g) $= -286\,kJ\,mol^{-1}$
$\Delta H_c^\ominus$ C$_4$H$_{10}$(g) $= -2878\,kJ\,mol^{-1}$

WORKED EXAMPLE

Questions that involve enthalpy change of combustion and enthalpy change of formation

Sometimes questions can check understanding of enthalpy change of combustion and enthalpy change of formation at the same time.

Determine the enthalpy change for the formation of propene, C_3H_6.

Use the enthalpy change of combustion values below:

$\Delta H_c^\ominus$ C(s) $= -394\,kJ\,mol^{-1}$
$\Delta H_c^\ominus$ H$_2$(g) $= -286\,kJ\,mol^{-1}$
$\Delta H_c^\ominus$ C$_3$H$_6$(g) $= -2058\,kJ\,mol^{-1}$

The equation for the enthalpy of formation of propene is:

$$3C(s) + 3H_2(g) \rightarrow C_3H_6(g)$$

Notice how it does not matter that you are working out the enthalpy change of formation of propene. The method is exactly the same because you have been given enthalpy change of combustion values.

A = +(3 × −394) + (3 × −286) − (−2058)
= 18 kJ mol⁻¹

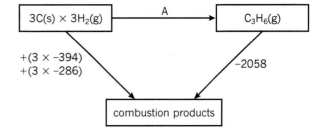

PRACTICE QUESTIONS

3 Determine the enthalpy change for the formation of but-2-ene, C_4H_8, using the enthalpy change of combustion values below:

$\Delta H_c^\ominus$ C(s) $= -394\,kJ\,mol^{-1}$
$\Delta H_c^\ominus$ H$_2$(g) $= -286\,kJ\,mol^{-1}$
$\Delta H_c^\ominus$ C$_4$H$_8$(g) $= -2705\,kJ\,mol^{-1}$

The equation for the enthalpy of formation of but-2-ene is:

$$4C(s) + 4H_2(g) \rightarrow C_4H_8(g)$$

4 Determine the enthalpy change for the formation of pent-1-ene, C_5H_{10}, using the enthalpy change of combustion values below:

$\Delta H_c^\ominus$ C(s) $= -394\,kJ\,mol^{-1}$
$\Delta H_c^\ominus$ H$_2$(g) $= -286\,kJ\,mol^{-1}$
$\Delta H_c^\ominus$ C$_5$H$_{10}$(l) $= -3509\,kJ\,mol^{-1}$

The equation for the enthalpy of formation of pent-1-ene is:

$$5C(s) + 5H_2(g) \rightarrow C_5H_{10}(l)$$

4.4 Hess's law 2

Calculating the enthalpy change for a reaction indirectly

Standard enthalpy change of formation, $\Delta H_f^{\ominus}$, values can be used in Hess's law calculations to determine the enthalpy change for a reaction that cannot be measured directly.

The standard enthalpy change of formation, $\Delta H_f^{\ominus}$, is the standard enthalpy change that takes place when one mole of a compound is formed from its constituent elements in their standard states under standard conditions.

Notice that this means that standard enthalpy change of formation for any element in its standard state is zero.

Notice how the arrows go from the elements to the compounds.

A is the enthalpy change for the reaction.

B is the enthalpy change of formation for the reactants.

C is the enthalpy change of formation for the products.

The direct route (A) is equal to the indirect route ($-B + C$).

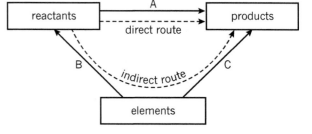

WORKED EXAMPLE

Calcium carbonate

Determine the enthalpy change for the decomposition of calcium carbonate.

$$CaCO_3(s) \rightarrow CaO(s) + CO_2(g)$$

Use the enthalpy change of formation values below:

$\Delta H_f^{\ominus}\ CaCO_3(s) = -1207\ kJ\,mol^{-1}$

$\Delta H_f^{\ominus}\ CaO(s)\ = -635\ kJ\,mol^{-1}$

$\Delta H_f^{\ominus}\ CO_2(g)\ = -394\ kJ\,mol^{-1}$

$A = -(-1207) + (-635) + (-394) = +178\ kJ\,mol^{-1}$

Notice that the positive sign for the enthalpy change indicates that it is an endothermic reaction.

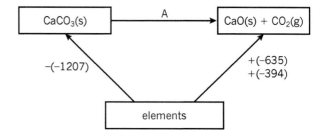

Potassium nitrate

Determine the enthalpy change for the decomposition of potassium nitrate, KNO_3.

$$2KNO_3(s) \rightarrow 2KNO_2(s) + O_2(g)$$

Use the enthalpy change of formation values below:

$\Delta H_f^{\ominus}\ KNO_3(s) = -495\ kJ\,mol^{-1}$

$\Delta H_f^{\ominus}\ KNO_2(s) = -370\ kJ\,mol^{-1}$

Notice that no value is given for oxygen as it is an element in its standard state.

First draw the Hess's law diagram for the reaction and add the enthalpy change of formation values for KNO_3 and KNO_2.

$A = -(2 \times -495) + (2 \times -370) = +250\ kJ\,mol^{-1}$

Notice that the positive sign for the enthalpy change indicates that it is an endothermic reaction.

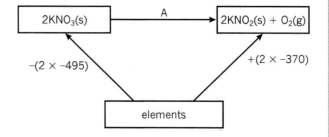

PRACTICE QUESTIONS

1   Determine the enthalpy change for the reaction:

$$CH_4(g) + 2H_2O(g) \rightarrow CO_2(g) + 4H_2(g)$$

Use these enthalpy change of formation values:

$\Delta H_f^\ominus$ $CH_4(g) = -75\,kJ\,mol^{-1}$

$\Delta H_f^\ominus$ $H_2O(g) = -242\,kJ\,mol^{-1}$

$\Delta H_f^\ominus$ $CO_2(g) = -394\,kJ\,mol^{-1}$

2   Determine the enthalpy change for the reaction:

$$C_2H_4(g) + 3O_2(g) \rightarrow 2CO_2(g) + 2H_2O(g)$$

Use these enthalpy change of formation values:

$\Delta H_f^\ominus$ $C_2H_4(g) = +52\,kJ\,mol^{-1}$

$\Delta H_f^\ominus$ $CO_2(g) = -394\,kJ\,mol^{-1}$

$\Delta H_f^\ominus$ $H_2O(g) = -242\,kJ\,mol^{-1}$

3   Determine the enthalpy change for the reaction:

$$CH_3OH(g) + 1\tfrac{1}{2}O_2(g) \rightarrow CO_2(g) + 2H_2O(l)$$

Use these enthalpy change of formation values:

$\Delta H_f^\ominus$ $CH_3OH(g) = -239\,kJ\,mol^{-1}$

$\Delta H_f^\ominus$ $CO_2(g) = -394\,kJ\,mol^{-1}$

$\Delta H_f^\ominus$ $H_2O(l) = -286\,kJ\,mol^{-1}$

4   Determine the enthalpy change for the reaction:

$$C_3H_6(g) + H_2(g) \rightarrow C_3H_8(g)$$

Use these enthalpy change of formation values:

$\Delta H_f^\ominus$ $C_3H_6(g) = -20\,kJ\,mol^{-1}$

$\Delta H_f^\ominus$ $C_3H_8(g) = -105\,kJ\,mol^{-1}$

5   Determine the enthalpy change for the reaction:

$$MgCO_3(s) \rightarrow MgO(s) + CO_2(g)$$

Use these enthalpy change of formation values:

$\Delta H_f^\ominus$ $MgCO_3(s) = -1096\,kJ\,mol^{-1}$

$\Delta H_f^\ominus$ $MgO(s) = -601\,kJ\,mol^{-1}$

$\Delta H_f^\ominus$ $CO_2(g) = -394\,kJ\,mol^{-1}$

6   Determine the enthalpy change for the reaction:

$$2NaNO_3(s) \rightarrow 2NaNO_2(s) + O_2(g)$$

Use these enthalpy change of formation values:

$\Delta H_f^\ominus$ $NaNO_3(s) = -425\,kJ\,mol^{-1}$

$\Delta H_f^\ominus$ $NaNO_2(s) = -359\,kJ\,mol^{-1}$

> **REMEMBER:** The enthalpy change of formation for $H_2O(g)$ is not the standard enthalpy change of formation for water. This is because the water is not in its standard state of a liquid but is instead a gas.

4.5 Bond enthalpy 1

Exothermic and endothermic reactions

Energy is released when bonds are made and required when bonds are broken. Bond enthalpy is the enthalpy change when one mole of a given bond in a gaseous molecule is broken by homolytic fission.

Sometimes the same bond, for example, C—H, can be found in many different molecules. The exact value of the bond enthalpy varies slightly from one molecule to another. The average bond enthalpy is the average enthalpy change for breaking one mole of the bond in a variety of gaseous molecules.

The enthalpy change for all exothermic reactions has a negative sign. More energy is released when new, stronger bonds are formed than was required to break the old, weaker bonds.

In contrast, the enthalpy change for all endothermic reactions has a positive sign. More energy is required to break the old, stronger bonds than is released when the new, weaker bonds are formed.

WORKED EXAMPLE

Calculating the enthalpy change for reactions

Formation of hydrogen chloride

Determine the enthalpy change for the reaction between hydrogen and chlorine to produce hydrogen chloride using the bond enthalpies shown in the table.

$H_2 + Cl_2 \rightarrow 2HCl$

Bond	Bond enthalpy / kJ mol^{-1}
H—H	+436
H—Cl	+432
Cl—Cl	+243

$$H—H \ + \ Cl—Cl \ \longrightarrow \ \begin{matrix} H—Cl \\ H—Cl \end{matrix}$$

The bonds broken in this reaction = $1 \times$ H—H = $1 \times 436 = 436$ kJ mol^{-1} and

$1 \times$ Cl—Cl = $1 \times 243 = 243$ kJ mol^{-1}

The total amount of energy required = $436 + 243 = 679$ kJ mol^{-1}

The bonds made in this reaction = $2 \times$ H—Cl = $2 \times 432 = 864$ kJ mol^{-1}

The total amount of energy released = 864 kJ mol^{-1}

So the enthalpy change for the reaction = total energy required − energy released

$= 679 − 864 = −185$ kJ mol^{-1}

Combustion of methane

Determine the enthalpy change for the complete combustion of methane using the bond enthalpies in the table below

Bond	Bond enthalpy / kJ mol^{-1}
O=O	+497
C=O	+805
O—H	+463
C—H	+413

During combustion, methane reacts with oxygen to form carbon dioxide and water.

$CH_4 + 2O_2 \rightarrow CO_2 + 2H_2O$

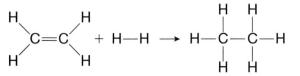

The bonds broken in this reaction = $4 \times$ C—H = $4 \times 413 = 1652$ kJ mol^{-1} and
$\qquad\qquad\qquad\qquad\qquad 2 \times$ O=O = $2 \times 497 = 994$ kJ mol^{-1}

The total amount of energy required = $1652 + 994 = 2646$ kJ mol^{-1}

The bonds made in this reaction = $2 \times$ C=O = $2 \times 805 = 1610$ kJ mol^{-1} and
$\qquad\qquad\qquad\qquad\qquad 4 \times$ O—H = $4 \times 463 = 1852$ kJ mol^{-1}

The total amount of energy released = $1610 + 1852 = 3462$ kJ mol^{-1}

So the enthalpy change for the reaction = total energy required − energy released
$\qquad\qquad\qquad\qquad\qquad\qquad = 2646 - 3462 = -816$ kJ mol^{-1}

? PRACTICE QUESTION

1   Use the bond enthalpy values in the table below to determine the enthalpy change for the following reactions.

a   The complete combustion of ethene.

$C_2H_4 + 3O_2 \rightarrow 2CO_2 + 2H_2O$

b   The reaction between ethene and hydrogen to produce ethane.

$C_2H_4 + H_2 \rightarrow C_2H_6$

c   The reaction between oxygen and hydrogen to produce water.

$O_2 + 2H_2 \rightarrow 2H_2O$

Bond	Bond enthalpy / kJ mol^{-1}
C—H	+413
C—C	+347
O=O	+497
C=O	+805
O—H	+463
H—H	+436
C=C	+612

REMEMBER: In these calculations the sign of the enthalpy change of reaction is significant. Exothermic reactions have a negative sign, whilst endothermic reactions have a positive sign. Always include the sign in your answers.

4.6 Bond enthalpy 2

Energy and bonds

Energy is released when bonds are made and required when bonds are broken.

An unknown bond enthalpy can be calculated if the enthalpy change for the reaction and the other bond enthalpies involved in the reaction are known.

WORKED EXAMPLE

Finding an unknown bond enthalpy value

Hydrogen bromide

A chemist adds hydrogen to bromine to produce hydrogen bromide.

$H_2 + Br_2 \rightarrow 2HBr$

$$H\!-\!H \; + \; Br\!-\!Br \; \longrightarrow \; \begin{array}{c} H\!-\!Br \\ H\!-\!Br \end{array}$$

The enthalpy change for the reaction is $-99\,kJ\,mol^{-1}$.

Determine the bond enthalpy value for the Br$-$Br bond using the bond enthalpy values shown in the table below.

Bond	Bond enthalpy / kJ mol^{-1}
H$-$H	+436
H$-$Br	+364

The bonds broken in this reaction = $1 \times$ H$-$H = 1×436 = $436\,kJ\,mol^{-1}$ and

$$1 \times Br\!-\!Br \; = \; ?$$

The total amount of energy required = $436 + $ Br$-$Br $kJ\,mol^{-1}$

The bonds made in this reaction = $2 \times$ H$-$Br = 2×364 = $728\,kJ\,mol^{-1}$

The total amount of energy released = $728\,kJ\,mol^{-1}$

So, the enthalpy change for the reaction = total energy required $-$ energy released

$$-99 = \quad (436 + Br\!-\!Br) \;\; - 728$$

So, Br$-$Br = $193\,kJ\,mol^{-1}$

Combustion of methane

A student burnt methane to produce carbon dioxide and water.

$CH_4 + 2O_2 \rightarrow CO_2 + 2H_2O$

$$\begin{array}{c} H \\ | \\ H\!-\!C\!-\!H \\ | \\ H \end{array} + \begin{array}{c} O\!=\!O \\ \\ O\!=\!O \end{array} \longrightarrow O\!=\!C\!=\!O \; + \; \begin{array}{c} H \diagdown O \diagup H \\ \\ H \diagdown O \diagup H \end{array}$$

The enthalpy change for the reaction is $-810\,kJ\,mol^{-1}$.

Determine the bond enthalpy value for the C$-$H bond using the bond enthalpy values shown in the table below.

Bond	Bond enthalpy / kJ mol^{-1}
O$=$O	+500
C$=$O	+805
O$-$H	+463

The bonds broken in this reaction = 4 × C—H

$$2 \times O{=}O = 2 \times 500 = 1000 \, kJ \, mol^{-1}$$

The total amount of energy required = 4 × C—H + 1000 kJ mol⁻¹

The bonds made in this reaction = 2 × C=O = 2 × 805 = 1610 kJ mol⁻¹

$$4 \times O{-}H = 4 \times 463 = 1852 \, kJ \, mol^{-1}$$

The total amount of energy released = 1610 + 1852 = 3462 kJ mol⁻¹

So, the enthalpy change for the reaction = total energy required − energy released

$$-810 = \quad 4 \times C{-}H + 1000 - 3462$$

4 × C—H = 1652 kJ mol⁻¹

So C—H = $\dfrac{1652}{4}$ = 413 kJ mol⁻¹

> **REMEMBER:** Experimental values for enthalpy changes may be slightly different to the theoretical values for enthalpy changes found using average bond enthalpy. The calculations use average bond enthalpies, which are average values for a particular bond in a variety of different gaseous molecules. The actual value for the bond in a given molecule may be slightly different.

PRACTICE QUESTIONS

1   Tyson adds hydrogen to chlorine to make hydrogen chloride.
    The reaction is shown below.
        $H_2 + Cl_2 \rightarrow 2HCl$
    The enthalpy change for the reaction is found to be −183 kJ mol⁻¹.
    Use the bond enthalpy values shown in the table below to determine the bond enthalpy value for the Cl—Cl bond.

Bond	Bond enthalpy / kJ mol⁻¹
H—H	+436
H—Cl	+431

2   Lucy adds hydrogen to ethene to make ethane.
    The reaction is shown below.
        $H_2 + C_2H_4 \rightarrow C_2H_6$
    The enthalpy change for the reaction is found to be −119 kJ mol⁻¹.
    Use the bond enthalpy values shown in the table below to determine the bond enthalpy value for the C—C bond.

Bond	Bond enthalpy / kJ mol⁻¹
C=C	+612
C—H	+410
H—H	+436

3   Janek adds hydrogen to iodine to make hydrogen iodide.
    The reaction is shown below.
        $H_2 + I_2 \rightarrow 2HI$
    The enthalpy change for the reaction is found to be −7 kJ mol⁻¹.
    Use the bond enthalpy values shown in the table below to determine the bond enthalpy value for the I—I bond.

Bond	Bond enthalpy / kJ mol⁻¹
H—H	+436
H—I	+297

5.1 Mathematical skills in chemistry practical work

Core mathematical skills

In order to be awarded a practical endorsement in A Level Chemistry, you need to complete a minimum of 12 practical assessments. A practical chemist must be proficient in standard form, significant figures, decimal places, SI units, and unit conversion.

Drawing graphs, percentages, and using algebra are mathematical skills that have been covered elsewhere in this book.

WORKED EXAMPLE
Using standard form

Standard form is writing a number in the format $A \times 10^x$ where A is a number from 1 to 10 and x is the number of places you move the decimal place.

For example, to express a large number such as $50\,000\,\text{mol}\,\text{dm}^{-3}$ in standard form, A = 5 and x = 4 as there are four numbers after the initial 5.

Therefore, it would be written as $5 \times 10^4\,\text{mol}\,\text{dm}^{-3}$.

To give a small number such as $0.000\,02\,\text{Nm}^2$ in standard form,

A = 2 and there are five numbers before it so x = –5.

So it is written as $2 \times 10^{-5}\,\text{Nm}^2$.

In science, very large and very small numbers are usually written in standard form. For example, Avagrado's constant 6×10^{23} and the mass of the electron 9×10^{-31}. If you reverse the process above you can write these numbers out in full, but it is more convenient to give them in standard form.

PRACTICE QUESTIONS

1   Give the following in standard form:

a	6000	b	400	c	80 000	d	9000
e	400 000	f	0.007	g	0.04	h	0.000 000 005
i	0.0234	j	0.000 002 3				

2   Give the following as ordinary numbers:

a	5.5×10^{-6}	b	6.5×10^{-8}	c	3.2×10^5	d	2.9×10^2
e	3.167×10^{-11}	f	1.115×10^4	g	1.412×10^{-3}	h	7.2×10^1
i	9.01×10^{-2}	j	1.17×10^6				

Significant figures and decimal places

In chemistry, you are often asked to express numbers to either three or four significant figures. The word significant means to 'have meaning'. A number that is expressed in significant figures will only have digits that are important to the number's precision.

For instance, in an experiment you find the relative atomic mass of lithium to be 6.9301. It is wrong to express the number to this many figures as it is beyond the precision of the equipment you used. You should express it to three significant figures.

For example, 6.9301 becomes 6.93 if written to three significant figures.

Likewise, 0.000 434 56 is 0.000 435 to three significant figures.

The first four digits have no bearing on the precision of the number and are not counted as significant. The last significant digit is beside a 5 or above, so just as with decimals, it must be rounded up.

All numbers after the first significant number are significant so, for example, 0.003 018 is 0.003 02 to three significant figures.

Sometimes numbers are expressed to a number of decimal places. The decimal point is a place holder and the number of digits afterwards is the number of decimal places.

For example, the mathematical number pi is 3 to zero decimal places, 3.1 to one decimal place, 3.14 to two decimal places, and 3.142 to three decimal places.

PRACTICE QUESTIONS

3   Give the following values in the stated number of decimal places (d.p.).

   a   4.763 (1 d.p.)   b   0.543 (2 d.p.)   c   12.89 (1 d.p.)
   d   2.956 (2 d.p.)   e   7.895 (2 d.p.)   f   1.998 (2 d.p.)
   g   1.005 (2 d.p.)   h   1.9996 (3 d.p.)

4   Give the following values in the stated number of significant figures (s.f.).

   a   36.937 (3 s.f.)   b   2.643 (2 s.f.)   c   19.6754 (4 s.f.)
   d   23 139 (3 s.f.)   e   258 (2 s.f.)   f   0.043 19 (2 s.f.)
   g   0.003 48 (2 s.f.)   h   7 999 032 (1 s.f.)

Unit conversions

Multiplication factor	Prefix	Symbol
$1\,000\,000\,000 = 10^9$	giga	G
$1\,000\,000 = 10^6$	mega	M
$1000 = 10^3$	kilo	k
$100 = 10^2$	hecto	h
$1 = 1$		
$0.01 = 10^{-2}$	centi	c
$0.001 = 10^{-3}$	milli	m
$0.000\,001 = 10^{-6}$	micro	μ
$0.000\,000\,001 = 10^{-9}$	nano	n

Unit conversions are common. For instance, you could be converting an enthalpy change of $488\,889\,\mathrm{J\,mol^{-1}}$ into $\mathrm{kJ\,mol^{-1}}$. A kilo is 10^3 so you need to divide by this number or move the decimal point three places to the left.

$$488\,889 \times 10^{-3}\,\mathrm{kJ\,mol^{-1}} = 488.889\,\mathrm{kJ\,mol^{-1}}$$

However, suppose you are converting from $\mathrm{mJ\,mol^{-1}}$ to $\mathrm{kJ\,mol^{-1}}$, so you need to go from 10^3 to 10^{-3}, or move the decimal point six places to the left.

$333\,\mathrm{mJ\,mol^{-1}}$ is $0.000\,333\,\mathrm{kJ\,mol^{-1}}$

Alternatively, if you want to convert from $333\,\mathrm{mJ\,mol^{-1}}$ to $\mathrm{nJ\,mol^{-1}}$, you would have to go from 10^{-9} to 10^{-3}, or move the decimal point six places to the right.

$333\,\mathrm{mJ\,mol^{-1}}$ is $333\,000\,000\,\mathrm{nJ\,mol^{-1}}$

PRACTICE QUESTION

5   Calculate the following unit conversions.

   a   15 kg to g   b   300 μm to m
   c   5 MJ to mJ   d   10 GW to MW
   e   10 GW to kW

5.2 Percentage yields

Actual and theoretical yields

Chemists can predict the amount of product made in a reaction.

The percentage yield links the actual amount of product made, in moles, and the theoretical yield, in moles:

$$\text{Percentage yield} = \frac{\text{actual amount (in moles) of product}}{\text{theoretical amount (in moles) of product}} \times 100\%$$

WORKED EXAMPLE

Calculating the theoretical yield

When calcium carbonate is heated it decomposes to form calcium oxide and carbon dioxide.

$$CaCO_3(s) \rightarrow CaO(s) + CO_2(g)$$

A chemist used 75.1 g of calcium carbonate, $CaCO_3$, in this reaction.

a Calculate the amount, in moles, of calcium carbonate that reacts.

$$n = \frac{m}{M} = \frac{75.1}{100.1} = 0.750 \, \text{mol}$$

b Calculate the amount, in moles, of calcium oxide that was made.

From the balanced equation:

the amount of calcium carbonate, in moles = amount of calcium oxide, in moles = 0.750 mol

c Calculate the mass, in grams, of calcium oxide made. Give your answer to three decimal places.

$$m = n \times M = 0.750 \times 56.1 = 42.1 \, \text{g}$$

The actual yield

However, in practice chemists often find that an experiment actually makes a smaller amount of product than expected.

This could be because:

- other reactions take place that produce an alternative product
- the reaction reaches an equilibrium and so does not go to completion
- some of the product is lost during transfer or purification, for example, during filtering or distillation.

WORKED EXAMPLE

Calculating the percentage yield

A student added ethanol to propanoic acid to make the ester, ethyl propanoate, and water.

$$C_2H_5OH + C_2H_5COOH \rightarrow C_2H_5COOC_2H_5 + H_2O$$

The experiment has a theoretical yield of 5.00 g.

The actual yield is 4.50 g.

The molar mass of $C_2H_5COOC_2H_5 = 102.0 \, \text{g mol}^{-1}$

Calculate the percentage yield of the reaction.

Actual amount of ethyl propanoate: $n = \frac{m}{M} = \frac{4.5}{102} = 0.0441 \, \text{mol}$

Theoretical amount of ethyl propanoate, $n = \frac{m}{M} = \frac{5.0}{102} = 0.0490 \, \text{mol}$

Percentage yield $= \frac{4.5}{5.0} \times 100\% = 90\%$

PRACTICE QUESTIONS

1 Calculate the percentage yield of a reaction that has a theoretical yield of 4.75 moles of product and an actual yield of 3.19 moles of product. Give your answer to three significant figures.

2 Calculate the percentage yield of a reaction that has a theoretical yield of 3.00 moles of product and an actual yield of 2.75 moles of product. Give your answer to three significant figures.

3 Calculate the percentage yield of a reaction with a theoretical yield of 12.00 moles of product and an actual yield of 6.25 moles of product. Give your answer to three significant figures.

Limiting reagents

In some reactions, one of the reactants is stated as being in excess. The other reactant is said to be the limiting reagent. The limiting reagent is the reactant that will be used up first and therefore cause the reaction to slow down and eventually stop.

WORKED EXAMPLE

An excess of magnesium is added to $50\,cm^3$ of $1.0\,mol\,dm^{-3}$ copper(II) sulfate solution.

$$Mg(s) + CuSO_4(aq) \rightarrow MgSO_4(aq) + Cu(s)$$

2.54 g of copper is produced.

Calculate the percentage yield of this reaction.

The amount, in moles, of $CuSO_4$ used, $n = \dfrac{V\,(\text{in }cm^3)}{1000} \times c = \dfrac{50}{1000} \times 1.0 = 0.05\,mol$

The theoretical amount, in moles, of copper produced in this reaction is also 0.05 mol.

The actual amount, in moles, of copper made $= \dfrac{m}{M} = \dfrac{2.54}{63.5} = 0.04\,mol$

Percentage yield $= \dfrac{0.04}{0.05} \times 100\% = 80\%$

PRACTICE QUESTIONS

4 An excess of zinc is added to $25.0\,cm^3$ of $1.0\,mol\,dm^{-3}$ iron(II) sulfate solution.

$$Zn(s) + FeSO_4(aq) \rightarrow ZnSO_4(aq) + Fe(s)$$

1.116 g of iron is produced.

Calculate the percentage yield of this reaction.

5 An excess of magnesium is added to $50.0\,cm^3$ of $1.0\,mol\,dm^{-3}$ aqueous hydrochloric acid solution.

$$Mg(s) + 2HCl(aq) \rightarrow MgCl_2(aq) + H_2(g)$$

0.953 g of magnesium chloride is produced.

Calculate the percentage yield of this reaction.

6 An excess of magnesium is added to $100.0\,cm^3$ of $0.5\,mol\,dm^{-3}$ iron(II) sulfate solution.

$$Mg(s) + FeSO_4(aq) \rightarrow MgSO_4(aq) + Fe(s)$$

0.558 g of iron is produced.

Calculate the percentage yield of this reaction.

Finding the reagent that is in excess

In some reactions the amount of both reactants (in moles) must be worked out. The balanced equation is then used to work out which of the reagents is in excess and which is the limiting reagent.

5.3 Atom economy

Wanted and unwanted products

Chemical reactions often produce more than one product. The product that is wanted is the desired product. The other (unwanted) products made in the reaction are called by-products.

Atom economy is a way of measuring the proportion of desired product compared with the total amount of product made in a reaction.

$$\% \text{ atom economy} = \frac{\text{molar mass of the desired product}}{\text{molar mass of all the product}} \times 100\%$$

Some by-products are of no use and have to be disposed of. This can be expensive and adds to the overall costs of the process. Chemists try to find uses for by-products. If the by-products can be sold on, it not only has economic benefits but it also means that natural resources are conserved for future use and the amount of waste for landfill sites is reduced.

WORKED EXAMPLE

Working out the % atom economy

Addition reactions

In addition reactions, reactants join together to form just one product.

Bromine is added to ethene to produce 1,2-dibromoethane.

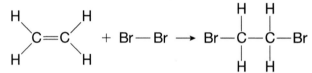

Determine the % atom economy of this reaction.

The molar mass of 1,2-dibromoethane = 187.8 g mol⁻¹

REMEMBER: All addition reactions have a % atom economy of 100%; all the reactant atoms are found in the desired product.

$$\% \text{ atom economy} = \frac{\text{molar mass of the desired product}}{\text{molar mass of all the products}} \times 100\% = \frac{187.8}{187.8} \times 100 = 100\%$$

PRACTICE QUESTIONS

1 Gordon added chlorine to ethene to produce 1,2-dichloroethane.

Determine the % atom economy of this reaction.
The molar mass of 1,2-dichloroethane = 99.0 g mol⁻¹

2 Katie adds bromine to propene to produce 1,2-dibromopropane.

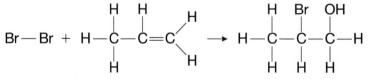

Determine the % atom economy of this reaction.
The molar mass of 1,2-dibromopropane = 201.8 g mol⁻¹

WORKED EXAMPLES

Further examples

1   Doug heated 2.5 g of calcium carbonate. It decomposed to form 1.2 g of the desired product, calcium oxide.

$$CaCO_3(s) \rightarrow CaO(s) + CO_2(g)$$

Determine the % atom economy of this reaction.

Give your answer to three significant figures.

Molar mass of CaO = 56.1 g mol^{-1}

Molar mass of CO$_2$ = 44.0 g mol^{-1}

% atom economy = $\dfrac{56.1}{100.1} \times 100 = 56.0\%$

2   Izzy decides to prepare potassium nitrite, KNO_2, by decomposing potassium nitrate, KNO_3. The equation for the reaction is:

$$2KNO_3 \rightarrow 2KNO_2 + O_2$$

Determine the % atom economy of this reaction.

Give your answer to three significant figures.

The molar mass of KNO_2 = 85.1 g mol^{-1}

The molar mass of O_2 = 32.0 g mol^{-1}

Total molar mass of KNO_2 in the reaction = 170.2

Molar mass of all the products = 202.2

% atom economy = $\dfrac{170.2}{202.2} \times 100 = 84.2\%$

PRACTICE QUESTIONS

3   Colin prepares a sample of the ester ethyl ethanoate by reacting ethanol and ethanoic acid.

$$C_2H_5OH + CH_3COOH \rightarrow CH_3COOC_2H_5 + H_2O$$

The molar mass of ethyl ethanoate = 88.0 g mol^{-1}

The molar mass of water = 18.0 g mol^{-1}

Determine the % atom economy of this reaction.

Give your answer to three significant figures.

4   Sara prepares a sample of propan-1-ol by reacting 1-chloropropane with sodium hydroxide solution.

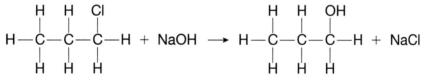

Determine the % atom economy of this reaction.

Give your answer to three significant figures.

5   Eric prepares a sample of ethanol by reacting sodium hydroxide solution with bromoethane.

$$NaOH + CH_3CH_2Br \rightarrow CH_3CH_2OH + NaBr$$

Determine the % atom economy of this reaction.

Give your answer three significant figures.

5.4 Percentage error in apparatus

Calculating percentage error

The percentage error of a measurement is calculated from the maximum error for the piece of apparatus being used and the value measured:

$$\text{Percentage error} = \frac{\text{maximum error}}{\text{measured value}} \times 100\%$$

WORKED EXAMPLE

Percentage error

An excess of zinc powder was added to $50\,cm^3$ of copper(II) sulfate solution to produce zinc sulfate and copper.

$$Zn(s) + CuSO_4(aq) \rightarrow ZnSO_4(aq) + Cu(s)$$

The initial and final temperature of the solution was recorded and used to work out the temperature change. Four students who completed the experiment recorded their results in the table below.

Student	Initial temperature/°C	Final temperature/°C	Temperature change/°C
A	21.0	24.9	3.9
B	21.2	25.0	3.8
C	21.0	26.5	5.5
D	21.1	25.1	4.0

a Suggest the advantage of comparing the results of four sets of experiments.

Allows you to spot any anomalous results.

b Calculate the mean temperature change for this reaction.

Result C appears to be anomalous so should not be included in the mean.

$$\text{The mean} = \frac{(3.9 + 3.8 + 4.0)}{3} = 3.9\,°C$$

c The measuring cylinder used to measure the copper(II) sulfate solution has a maximum error of $\pm 2\,cm^3$.

Calculate the percentage error.

$$\text{Percentage error} = \frac{2}{50} \times 100\% = 4\%$$

d A thermometer has a maximum error of $\pm 0.05\,°C$.

  i Calculate the percentage error when the thermometer is used to record a temperature of 25.0 °C.

$$\text{Percentage error} = \frac{0.05}{25.0} \times 100\% = 0.2\%$$

  ii Calculate the percentage error when the thermometer is used to record a temperature rise of 3.9 °C. Give your answer to three significant figures.

$$\text{Percentage error} = \frac{(2 \times 0.05)}{3.9} \times 100\% = 2.56\%$$

Notice that two temperatures are required to calculate the temperature change so the maximum error is doubled.

PRACTICE QUESTIONS

1   A measuring cylinder has a maximum error of $\pm 1\,cm^3$. Calculate the maximum error when recording these values. Give your answers to three significant figures.

   a $25.0\,cm^3$ b $80.0\,cm^3$ c $38.0\,cm^3$

2   A thermometer has a maximum error of $\pm 0.5\,°C$. Calculate the maximum error when recording these values. Give your answers to three significant figures.

   a $10.0\,°C$ b $15.0\,°C$ c $83.0\,°C$

3   A gas syringe has a maximum error of $\pm 0.5\,cm^3$. Calculate the maximum error when recording these values. Give your answers to three significant figures.

   a $21.0\,cm^3$ b $26.0\,cm^3$ c $43.0\,cm^3$

4   A thermometer has a maximum error of $\pm 0.5\,°C$. Calculate the maximum error when recording these temperature rises. Give your answers to three significant figures.

   a $12.0\,°C$ b $21.0\,°C$ c $37.6\,°C$

WORKED EXAMPLE

✓

Percentage error in titrations

A chemist carries out a titration between sodium hydroxide solution and hydrochloric acid.

$$HCl(aq) + NaOH(aq) \rightarrow NaCl\,(aq) + H_2O(l)$$

A pipette is used to transfer $25.0\,cm^3$ of sodium hydroxide solution to a flask. The pipette has a maximum error of $\pm 0.5\,cm^3$.

The results for the amount of hydrochloric acid required in the titration are shown below. The burette has a maximum error of $\pm 0.05\,cm^3$.

Experiment	Volume of hydrochloric acid / cm^3
1	22.20
2	21.80
3	21.70

a   Calculate the mean volume of hydrochloric acid used.

   Result 1 appears to be anomalous so should not be included in the mean.

   The mean $= \dfrac{(21.80 + 21.70)}{2} = 21.75\,cm^3$

b   Calculate the maximum percentage error in the measurement of the volume of hydrochloric acid. Give your answer to three significant figures.

   Two volume readings are required to calculate the volume of hydrochloric acid added from the burette. Calculating the percentage error in the volume of hydrochloric acid used in the first experiment:

   Percentage error $= \dfrac{(2 \times 0.05)}{22.2} \times 100 = 0.450\%$

PRACTICE QUESTION

5   The experiment above was repeated by another group and the new results are shown opposite.

   a   Calculate the mean volume of hydrochloric acid used.

   b   Calculate the maximum error in the measurement of the volume of hydrochloric acid in Experiment 1. Give your answer to three significant figures.

Experiment	Volume of hydrochloric acid / cm^3
1	21.60
2	21.70
3	21.30

5.5 Mass spectrometry and relative atomic mass

Calculating the mean masses of isotopes in order to find the relative atomic mass, A_r

Mass spectrometry is useful because it enables chemists to find the relative abundance of isotopes, atomic and molecular mass, and information regarding the structure of a molecule. They can do this with only a small amount of sample.

In order to achieve this, the sample is first injected and ionised. It is then sorted and separated according to mass and charge by external electric and magnetic fields.

The ions are then detected and results are displayed on a screen.

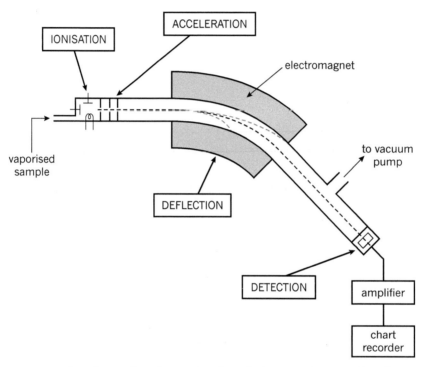

This mass spectrum can then be used to determine the relative atomic mass, A_r, of an elemental sample.

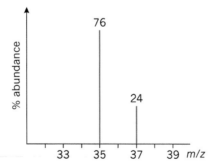

A sample of the element chlorine is injected and ionised. From this, you can work out the relative atomic mass from the percentage abundance of all the naturally occurring isotopes.

WORKED EXAMPLE

How to calculate the relative atomic mass of chlorine

$$\text{Relative atomic mass, } A_r = \frac{\sum(\text{mass of isotope} \times \% \text{ abundance})}{100}$$

From the mass spectrum, chlorine has two naturally occurring isotopes:

^{35}Cl with a mass abundance of 76%

^{37}Cl with a mass abundance of 24%

$$A_r \text{ of Cl} = \frac{(35 \times 76) + (37 \times 24)}{100} = 35.5 \text{ (1 d.p.)}$$

> **REMEMBER:** Relative atomic mass is usually given to 1 d.p. or 3 s.f. Make sure you write the correct number of significant figures or decimal places.
>
> Make sure your total % abundance adds up to 100%.

PRACTICE QUESTIONS

1. Use the table below to calculate the relative atomic mass of strontium from its mass spectrum data.

Mass number	% Abundance
84	0.6
86	9.9
88	89.5

2. Calculate the A_r of magnesium if the isotope masses and abundances are:
^{24}Mg = 78%, ^{25}Mg = 10%, ^{26}Mg = 12%.

3. Calculate the A_r of calcium from the following isotopic abundances:
^{40}Ca = 97%, ^{42}Ca = 0.65%, ^{44}Ca = 2.35%

4. Calculate the A_r of iron from the following isotopic abundances:
^{54}Fe = 5.0%, ^{56}Fe = 92.0%, ^{57}Fe = 2.4%, ^{58}Fe = 0.6%

STRETCH YOURSELF!

The relative atomic mass of an element is the weighted mean atomic mass of the element relative to $\frac{1}{12}$ of the mass of an atom of carbon-12.

a. Explain the meaning of 'weighted mean' in the definition.

b. Why use $\frac{1}{12}$ of the mass of an atom of carbon-12?

5.6 Mass spectroscopy (fragmentation of organic compounds)

Interpreting numerical data to investigate molecular structure

As well as finding the relative atomic mass of an element, mass spectrometry can also be used to find the relative formula mass by finding the molecular ion, M^+. It can also provide data to enable the molecular structure of the compound to be deduced from the fragmentation patterns from smaller ions.

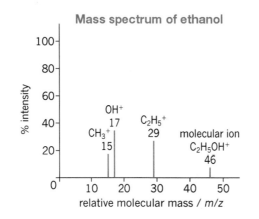

Mass spectrum of ethanol

> REMEMBER: The molecular ion is formed when one electron is removed from the molecule. The molecular ion is the largest abundant ion. Larger ions with a small relative abundance are due to isotopic masses and should not be confused with the molecular ion.

WORKED EXAMPLE

Interpretation of the mass spectrum of ethanol, CH_3CH_2OH

Looking at the spectrum above, the molecular ion is
$CH_3CH_2OH^+ = 46\,m/z$

Smaller fragments can be observed;

$CH_3CH_2^+ = 29\,m/z$

$OH^+ = 17\,m/z$

$CH_3^+ = 15\,m/z$

PRACTICE QUESTIONS

Identification of structure through mass spectra

For each of the following problems:
- use graph paper to draw an accurate diagram of the mass spectrometer data as it would appear in a chart
- use the data to identify the molecular ion
- identify the smaller fragementation ions on the chart and deduce the structure.

1 A compound with C_3H_7Cl has abundant ions at 15.

2 A compound with $C_5H_{10}O$ has abundant ions at 29, 57, and 86.

3 Another compound with C_5H_{12} has abundant ions at 29, 43, 57, and 72.

STRETCH YOURSELF!

A compound with two molecular ions

a Explain why two molecular ion peaks are found in the mass spectrum of chloroethane at *m/z* values of 64 and 66, with relative intensities in the ratio of approximately 3 to 1 respectively.

b Write an equation for the fragmentation of $CH_3CH_2Cl^+$ giving rise to a peak at *m/z* value of 29.

PRACTICE QUESTIONS

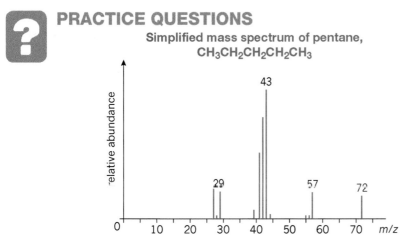

Simplified mass spectrum of pentane, $CH_3CH_2CH_2CH_2CH_3$

4 Identify the fragmentation peaks for pentane.

   a 29 *m/z*

   b 43 *m/z*

   c 57 *m/z*

   d 72 *m/z*

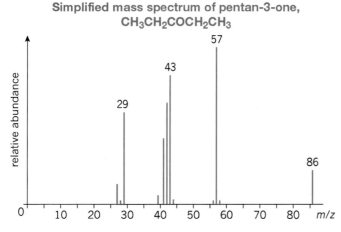

Simplified mass spectrum of pentan-3-one, $CH_3CH_2COCH_2CH_3$

5 Identify the fragmentation peaks for pentanone.

   a 29 *m/z*

   b 57 *m/z*

   c 86 *m/z*

5.7 NMR

Carbon-13 and proton nuclear magnetic resonance

NMR is a technique used to determine the structure of organic compounds. You are expected to interpret two types of NMR at A Level: carbon-13 (^{13}C) NMR and high-resolution proton (^1H) NMR. Both types of NMR provide information about the position of atoms and their neighbouring atoms (environment) in a molecule.

By interpreting NMR data, you should be able to predict the number of environments from the number of peaks and determine the type of environment using a chemical shift table.

In ^1H NMR you will also need to interpret splitting patterns by using the $(n + 1)$ rule to suggest the numbers of protons in adjacent environments. As such, using ^1H NMR enables a more complete identification of the structure.

> **REMEMBER:** Don't get your NMR confused!
> In ^{13}C NMR propan-2-ol has two carbon environments.
> In NMR it has three proton environments.

✓ WORKED EXAMPLE

Carbon-13 NMR

Interpret the following ^{13}C NMR spectra of propan-1-ol and propan-2-ol

1 Establish the number of carbon environments.
   In propan-1-ol there are three non-equivalent carbon environments. To determine whether a carbon atom is in a non-equivalent environment, the adjacent carbon atom must be considered. There are two CH_2 groups in propan-1-ol, but one of the CH_2 group is in an environment where an OH is adjacent and the other is in environment adjacent to a CH_3 group. They are not equivalent and therefore have to be treated as two separate environments.
   In propan-2-ol as both CH_3 groups are adjacent to a CH group, both CH_3s are equivalent.
2 Use the ^{13}C NMR shift table to identify the peaks.
3 The height of each peak indicates the ratio of the number of carbons in the environment. You can see that there are equal numbers of carbons in each environment in propan-1-ol by the height of the peaks. In propan-2-ol you can see that there are two CH_3 groups and that this peak is double the height of the CH peak. The height of peaks will be indicated in the question, so you will not need to measure them.

? PRACTICE QUESTION

1 State the number of carbon environments in the following compounds.

   a butanoic acid   b butan-1,4-diol   c propyl ethanoate

   d benzene   e phenol   f 1,3-dinitrobenzene

Aromatic carbon compounds

With an aromatic carbon compound, such as methylbenzene, it is best to draw a line of symmetry to work out the number of carbon environments.

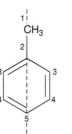

> By drawing out the molecule and drawing the line of symmetry, it is possible to see that there are five carbon environments.

✓ WORKED EXAMPLE
High resolution proton ¹H NMR

Interpret the high-resolution proton spectrum of 2-chloropropane

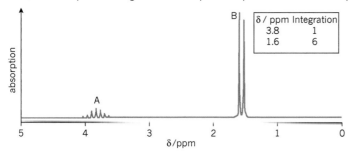

δ / ppm	Integration
3.8	1
1.6	6

> NOTE: In an exam, you will not see true NMR spectra – they are drawings! Even the tiniest peak is intentional. There are 7 peaks here!

1   Establish the number of proton environments. The number of proton environments can be determined by the number of independent peaks. The presence of two peaks above indicates two proton environments.

2   Use the proton NMR shift table to identify likely environments.

3   Calculate the ratio of protons in each environment. The area of peak B is six times the size of peak A. So for every one proton in environment A, there are six protons in environment B.

4   The NMR signal of a proton will be split into $n + 1$ peaks of the same height, where n is the number of hydrogens on an adjacent carbon. Deduce the number of protons in neighbouring environments from A and B using the $(n + 1)$ rule. B is a doublet splitting pattern of $(n + 1 = 2)$ which means the adjacent carbon atom is bonded to one proton. A is a septet splitting pattern $(n + 1 = 7)$, which means there are six protons in neighbouring environments.

5   It is now possible to deduce the structure of 2-chloropropane:

$A = CH_3$
$B = -CH-$

> NOTE: Protons bonded to nitrogen and oxygen atoms do not produce splitting patterns.

? PRACTICE QUESTIONS

2

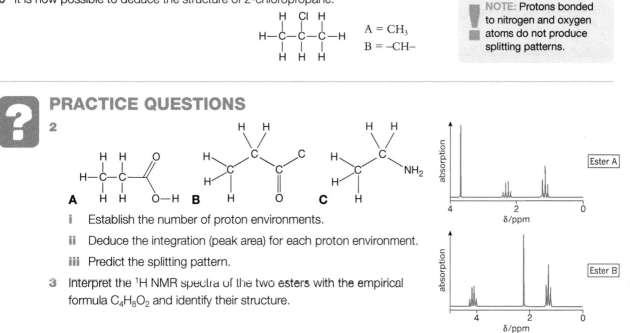

A   **B**   **C**

   i   Establish the number of proton environments.

   ii   Deduce the integration (peak area) for each proton environment.

   iii   Predict the splitting pattern.

3   Interpret the ¹H NMR spectra of the two esters with the empirical formula $C_4H_8O_2$ and identify their structure.

5.8 Chromatography and combined techniques

Chromatography and the calculation of R_f values

All chromatography works on the principle of separating of components between the mobile phase (solvent) and the stationary phase (material the solvent travels through). In order to account for variations in width and height of the stationary phase, you can calculate an R_f value. The R_f of a substance is the ratio of the distance moved by the component along the stationary phase, to the total distance travelled by the mobile phase. The R_f value of a component is always the same, and can be referenced against a known value – leading to identification.

WORKED EXAMPLE

Calculating the R_f value of a component

$R_f = 0.89$

$R_f = 0.44$

$R_f = ?$

y

x

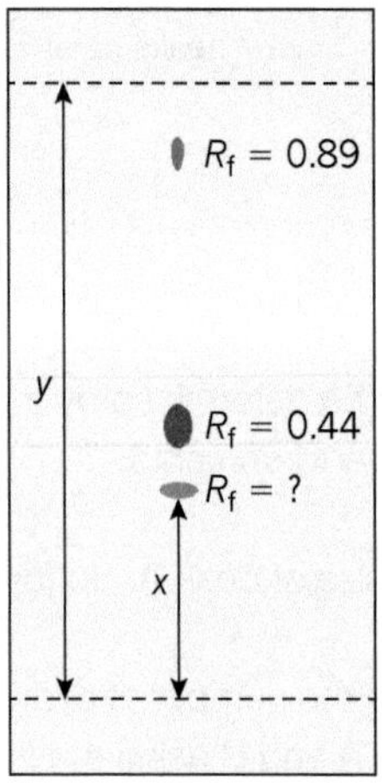

$R_f = x/y$

To calculate the R_f value $= \dfrac{\text{distance component}}{\text{height of mobile phase.}}$

Birute measures the distance, x, moved by a component, and how far the mobile phase travelled, y, on the chromatogram.
She found that $x = 48\,mm$ and $y = 141\,mm$.

Calculate the R_f value of the component.

$R_f = \dfrac{x}{y}$

$R_f = \dfrac{48}{141}$

$R_f = 0.34$

PRACTICE QUESTIONS

1 The table below shows the distance that six components travelled in a mobile phase

Substance	Distance travelled / cm
methanol solvent (mobile phase)	10
A	6
B	3
C	4
D	0

a Calculate the R_f for components A–D.

b Suggest what you can conclude about component D.

c Predict what would happen to the R_f values of A–D if the chromatogram was repeated with a less polar solvent.

Combination of chromatography with other techniques

Chromatography is often combined with other analytical techniques, such as mass spectrometry, infrared spectroscopy, and nuclear magnetic resonance. The advantage of chromatography over the other techniques mentioned is that it can separate components of mixtures prior to using the other instrumental techniques.

In order to solve the problem below, you will have to be familiar with empirical formulae, mass spectrometry, infrared, and proton NMR, which are covered earlier in this book.

STRETCH YOURSELF!

- A mixture of two types of organic compunds, X and Y, is separated by chromatography and then analysed using a mass spectrometer. Both have a molecular ion of 86 m/z.

- Neither X nor Y can be oxidised by heating under reflux with acidified $K_2Cr_2O_7$. However, both of the molecules can be reduced with $NaBH_4$.

- Combustion analysis revealed that both organic compounds contained 69.8% carbon, 11.6% hydrogen, and 18.6% oxygen by mass.

- The mass spectrum of compound X gave peaks at 15, 29, 43, and 57. The mass spectrum of compound Y gave peaks at 15, 29, and 57.

- The infra red spectra of both X and Y gave sharp peaks at 2700 cm⁻¹ and 1800 cm⁻¹.

Proton NMR of compound X

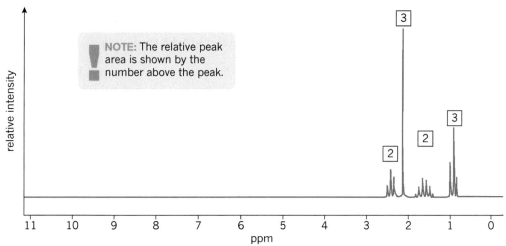

Proton NMR of compound Y

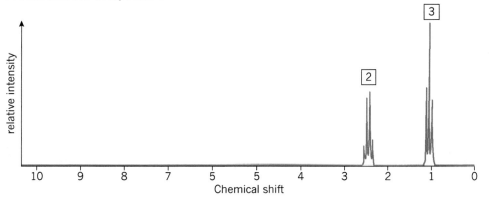

Use the information provided to identify the structures of compounds X and Y.

5.9 Entropy

Calculating the entropy change for a reaction given the entropies of reactants and products

Entropy, S, is the quantitative measurement of the degree of disorder of a system. A system becomes more energetically stable the more disordered it is.

A good analogy to this is to consider a plate that is dropped on the floor – it will smash into many pieces. Yet if the pieces are picked up and dropped again they do not reassemble. This is because an isolated system drives to become more disordered. In the same way, gas molecules spread out from a source to fill a space – increasing their entropy.

In any chemical reaction, the entropy of the chemicals changes and so does the entropy of the surroundings. Just as when the a plate is smashed, the total entropy increases. The total entropy, ΔS_{total}, must also increase for a chemical reaction to be feasible.

$$\Delta S_{total} = \Delta S_{system} + \Delta S_{surroundings}$$

$$\Delta S_{surroundings} = -\frac{-\Delta H^{\ominus}}{T}$$

$\Delta H^{\ominus}$ is enthalpy in $J\,mol^{-1}$ and T is the temperature in K.

Standard molar entropies are entropy changes that occur under standard conditions and are measured in joules per kelvin per mole: $J\,K^{-1}\,mol^{-1}$.

These values are given in an exam paper and can be used to calculate the total entropy changes and predict the feasibility of reaction.

$$\Delta S_{system} = \sum S^{\ominus}_{products} - \sum S^{\ominus}_{reactants}$$

WORKED EXAMPLE

The entropy change in the condensation of water at 298 K

$$H_2O(g) \rightarrow H_2O(l) \qquad \Delta H^{\ominus} = -44\,kJ\,mol^{-1}$$

Water vapour can sometimes spontaneously condense to water at room temperature.

Standard entropy, $S^{\ominus}$, of:

$H_2O(g) = 188.7\,J\,K^{-1}\,mol^{-1}$

$H_2O(l) = 70.0\,J\,K^{-1}\,mol^{-1}$

$\Delta S_{system} = \sum S^{\ominus}_{products} - \sum S^{\ominus}_{reactants}$

$\Delta S^{\ominus}_{system} = 70 - 188.7 = -118.7\,J\,K^{-1}\,mol^{-1}$

This reaction would seem unfeasible because it has a negative entropy change. However, first you must consider the entropy of the surroundings.

$$\Delta S_{surroundings} = \frac{-\Delta H^{\ominus}}{T}$$

$\Delta H = -44\,kJ\,mol^{-1}$ so $\Delta H^{\ominus} = -44 \times 1000 = -44\,000\,J\,mol^{-1}$

$$\Delta S_{surroundings} = -\frac{-44\,000}{298} = 147.65\,J\,K^{-1}\,mol^{-1}$$

$\Delta S_{total} = \Delta S_{system} + \Delta S_{surroundings}$

$\Delta S_{total} = -118.7 + 147.65 = +28.95\,J\,K^{-1}\,mol^{-1}$

The total entropy change is positive, so the reaction must be feasible.

> **REMEMBER:** Gases must have higher entropies than liquids and solids.
>
> Ions and molecules in solution must have higher entropies than solids.
>
> A large molecule that breaks down into smaller molecules must increase in entropy.
>
> Total entropy must always increase, even though the system entropy may not.

PRACTICE QUESTIONS

1 Predict whether ΔS_{system} of the following will be positive or negative.

 a $CuCO_3(s) \rightarrow CuO(s) + CO_2(g)$

 b $N_2(g) + 3H_2(g) \rightarrow 2NH_3(g)$

2 The combustion of magnesium and the change in enthalpy are given by:

$$2Mg(s) + O_2(g) \rightarrow 2MgO(s) \qquad \Delta H^{\ominus} = -1204\,kJ\,mol^{-1}$$

The standard entropy values are:
$S^{\ominus}(Mg)\ \ = 32.71\,J\,K^{-1}\,mol^{-1}$
$S^{\ominus}(O_2)\ \ = 204.9\,J\,K^{-1}\,mol^{-1}$
$S^{\ominus}(MgO) = 26.8\,J\,K^{-1}\,mol^{-1}$

 a Calculate the standard entropy of the reactants.
 Remember to take into account number of moles.

 b Calculate the standard entropy of the products.

 c What is the standard entropy change of the system?

 d Calculate ΔS of the surroundings at 298 K.
 Remember H must be converted to joules.

 e Calculate the total entropy and state whether the reaction is feasible.

STRETCH YOURSELF!

The standard entropies of noble gases

The standard entropies of noble gases increase from helium to xenon.

 a Describe the change in atomic structure down Group 0.

 b Explain how this affects the van der Waals forces of attraction.

 c i Describe the trend in boiling points down Group 0.
  ii Suggest how the trend affects the enthalpy change of vaporisation.

 d i Explain why the standard entropies of noble gases increase from helium to xenon.
  ii Explain why the trend is less significant at lower temperatures.

5.10 Gibbs free energy

Rearranging equations, substituting data, and solving equations to find the Gibbs free energy of a reaction

The likelihood of a reaction occurring is dependent on the entropy, $\triangle S$, the change in enthalpy, $\triangle H$, and the temperature, T.

Working with entropy values to predict the feasibility of a reaction is awkward because the entropy values of the system and the surroudings have to be accounted for. However,

$$\triangle S_{total} = \triangle S_{system} + \triangle S_{surroundings}$$

can be converted to:

$$\triangle G = \triangle H - T\triangle S_{system}$$

where $\triangle G$ is the Gibbs free energy measured in $J\,mol^{-1}$.

A negative value for $\triangle G$ means the reaction is energetically feasible, or spontaneous. It will happen by itself; it does not need an input of energy to occur. Gibbs free energy does not tell you about the activation energy or the rate of reaction, so the reaction may be so slow it appears not to happen.

WORKED EXAMPLE

Calculating the Gibbs free energy

Consider the reaction of ammonia and hydrogen chloride at 298 K:

$NH_3(g) + HCl(g) \rightarrow NH_4Cl(s)$

The enthalpy change of the reaction

$$\triangle H^{\ominus} = -315\,kJ\,mol^{-1} = -315 \times 1000 = -315\,000\,J\,mol^{-1}$$

$$\triangle S^{\ominus} = -284.5\,J\,K^{-1}\,mol^{-1}$$

Using

$$\triangle G = \triangle H - T\triangle S_{system}$$

$$\triangle G = -315\,000 - (298 \times 284.5)$$

$$= -315\,000 - 84\,781$$

$$= -399\,781\,J\,mol^{-1}$$

> **REMEMBER:** For most reactions $\triangle G$ and $\triangle H$ are similar because of a negligible $-T\triangle S_{system}$ value. Exothermic reactions are usually stated to be feasible except at large temperatures when $-T\triangle S_{system}$ is significant.

The reaction has the negative standard enthalpy change of reaction expected because 2 moles of gas are forming 1 mole of a solid.

The large $\triangle G$ and $\triangle H$ show the reaction to be feasible. It will happen by itself without an input of energy

---

PRACTICE QUESTIONS

1   For each equation below, calculate $\triangle G$ using $\triangle G = \triangle H - T\triangle S_{system}$ Also, state whether or not the reaction will be feasible. The temperature given is 298 K.

a $CH_3OH(l) + 1\frac{1}{2}O_2(g) \rightarrow CO_2(g) + 2H_2O(g)$ $\triangle H = -638.4\,kJ\,mol^{-1}$
$\triangle S = 156.9\,J\,K^{-1}\,mol^{-1}$

b $2NO_2(g) \rightarrow N_2O_4(g)$ $\triangle H = -57.2\,kJ\,mol^{-1}$
$\triangle S = -175.9\,J\,K^{-1}\,mol^{-1}$

2 Calculate $\triangle G$ for the following reaction at 298 K.

$$CaO(s) + H_2O(l) \rightarrow Ca(OH)_2(s)$$

	$\Delta H_f / kJ\,mol^{-1}$	$S / J\,K^{-1}\,mol^{-1}$
H_2	0.0	131.0
CaO	−635.5	39.7
$H_2O(l)$	−285.9	70.0
$Ca(OH)_2(s)$	−986.9	76.0
CH_3OH	−201.0	240.0
C_2H_5OH	−235.0	161.0
CO	−110.0	198.0
CO_2	−393.0	214.0
O_2	0.0	205.0

> **REMEMBER:** ΔH_r is the enthalpy change of reaction. It can be calculated by subtracting standard enthalpy of formation of products from the standard enthalpy of formation of reactants according to Hess's law.

a First find $\triangle H$ $\Delta H_r = \Delta H_f \text{ (products)} - \Delta H_f \text{ (reactants)}$

b Find $\triangle S$ $\Delta S = S_{products} - S_{reactants}$

c Find $\triangle G$ $\Delta G = \Delta H - T\Delta S$

3 Calculate $\triangle G$ for the following reaction at 298 K.

$$C_2H_5OH + 3O_2 \rightarrow 2CO_2 + 3H_2O$$

STRETCH YOURSELF!

When will this reaction cease to be feasible?

$CO(g) + 2H_2(g) \rightarrow CH_3OH(g)$

Using the table above, calculate $\triangle S$ and $\triangle H$. Then determine the temperature at which the reaction ceases to be feasible or spontaneous.

6.1 Boltzmann distributions

The distribution of energies in a gaseous sample

Chemists can use Boltzmann distributions to show the energies of the particles in a sample of gas at a particular temperature.

These diagrams show that most particles have an energy that falls into a narrow band, but a few particles have much more energy and a few have much less energy.

The shape of the curve is not symmetrical.

The line starts at the origin, showing that none of the particles have zero energy. The line approaches the x-axis, but does not cross it.

At higher temperatures, the peak of the line is lower and shifts to the right.

Only the particles that have more energy than the activation energy can react.

The area under the line represents the total number of particles. Although the shape of the curve changes at different temperatures, the area under the graph is constant.

WORKED EXAMPLE

Temperature and Boltzmann distributions

The figure below shows the distribution of energies in a sample of gas at 273 K.

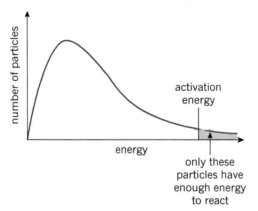

Add a line to the diagram to show the distribution of energies in the sample of gas at 298 K. Explain how the diagram shows that increasing the temperature of the sample increases the rate of reaction.

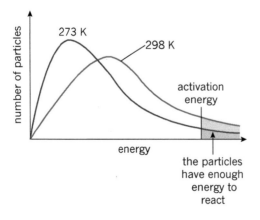

At a higher temperature, the particles have more kinetic energy and the peak of the curve is shifted towards the higher energy (right-hand) side. This means that more of the particles have energy greater than the activation energy, so there are more successful collisions and a faster rate of reaction.

The peak is lower at a higher temperature because the total area under the line represents the total number of particles, which remains the same.

PRACTICE QUESTION

1   The diagram below shows the distribution of energies in a sample of gas at 100°C.

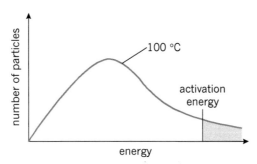

Add a line to show the distribution of particles at 50°C. Explain why the rate of reaction is slower at a lower temperature.

WORKED EXAMPLE

Catalysts and Boltzmann distribution

Adding a catalyst does not change the shape of the distribution curve but it does change the activation energy.

Explain how a catalyst increases the rate of reaction.

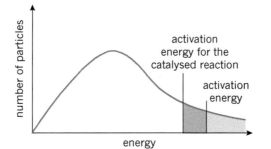

Adding a catalyst decreases the activation energy. This means that more particles have enough energy to react, so there are more successful collisions and the rate of reaction increases.

PRACTICE QUESTION

2   Zeeshan carries out a reaction between gaseous particles using a catalyst. Explain how removing the catalyst will decrease the rate of the reaction. Include a Boltzmann distribution diagram in your answer.

6.2 Concentration–time graphs

Deducing rate and order of a reaction from a graph

When investigating the reaction rate during a reaction, you could measure a volume of gas or colour change to work out the concentration of a reactant during the experiment. By measuring this concentration at repeated intervals you could plot a concentration–rate graph. By plotting a tangent or measuring the gradient you can calculate the rate. The order of the reaction can also be calculated.

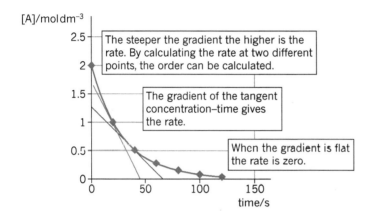

The steeper the gradient the higher is the rate. By calculating the rate at two different points, the order can be calculated.

The gradient of the tangent concentration–time gives the rate.

When the gradient is flat the rate is zero.

WORKED EXAMPLE

Looking at the graph above. When the concentration of A has halved to $1.0\,mol\,dm^{-3}$, the tangent intercepts the y-axis at 1.75 and the x-axis at 48.

The gradient is $\dfrac{-1.75}{48} = -0.0365$ (3 s.f.).

So the rate is $0.0365\,mol\,dm^{-3}\,s^{-1}$.

When the concentration of A halves again to $0.5\,mol\,dm^{-3}$,

the gradient is $\dfrac{-1.25}{65} = -0.0192$

So the rate is $0.0192\,mol\,dm^{-3}\,s^{-1}$.

Allowing for a small experimental error, when the concentration is halved the rate is halved.

Conversely, if you double the concentration of A the rate doubles. The order with respect to reactant A is said to be 1.

PRACTICE QUESTIONS

1 Plot a concentration–time graph for the data in this table.

CH$_3$Br / mol dm^{-3}	0.0100	0.0064	0.0042	0.0027	0.0018	0.0012	0.0008	0.0005
Time / min	0	5	10	15	20	25	30	35

> **REMEMBER:** Work out a suitable scale and range before you start plotting your graph. If the scale is too small it will make it difficult to calculate the rate.

2 Calculate the rate of the reaction at 10 minutes and 20 minutes.

3 Estimate the order of the reaction.

STRETCH YOURSELF!

Plotting a rate graph

a Find the gradient at five points along the concentration–time graph. This gives you five rate points for a rate–concentration graph.

Plot the points to identify the order of reaction. The graphs below show zero order, first order, and second order graphs.

Draw a curve through the points and identify the order of the reaction.

> **REMEMBER:**
> **Zero order** – If you double the reactant the rate stays the same.
> **First order** – If you double the reactant the rate doubles.
> **Second order** – If you double the reactant the rate quadruples.

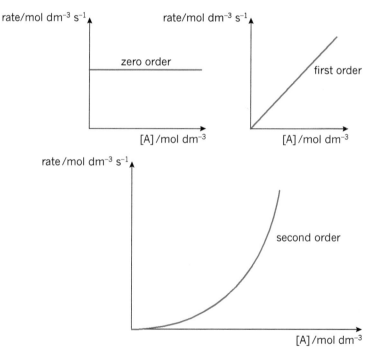

The following results were obtained from the isomerisation reaction of butane to cyclobutane.

Time / hours	0	2	5	10	20	30
% of cyclobutane remaining	100	91	79	63	40	25

b Write an equation to show the reaction involved.

c Draw a concentration–time graph.

d Using tangents, convert to a rate–concentration graph and identify the order of the reaction.

6.3 Half-life and initial rates

Deducing the half-life of a first-order reaction from a concentration-time graph, and using an initial rate graph to calculate the order of a reaction

The half-life of a reactant is the time taken for the concentration of the reactant to fall by half.

In chemistry, half-life is not exclusively concerned with radioactivity but could be concerned with the decomposition of a reactant, for example.

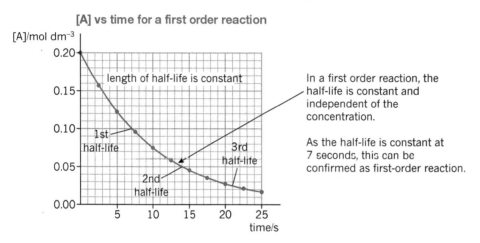

As discussed in the Stretch Yourself! activity in Topic 6.2, one way of deducing the order of a reaction is recording the changing concentration of a reactant over time and then plotting a rate–concentration graph.

The order of a reaction can also be calculated by changing the initial concentration of the reactant and measuring the initial rates. This has the advantage that the order of reaction with respect to reactants can be calculated without a graph. This is done by changing the concentration of one of the reactants and keeping the concentration of the others constant.

WORKED EXAMPLE

Calculating the order of a reaction using the initial rate method

Experiment	$[H_2(g)]$ initial $/mol\,dm^{-3}$	$[I_2(g)]$ initial $/mol\,dm$	Initial rate $/mol\,dm^{-3}\,s^{-1}$
1	0.01	0.01	4.70×10^{-6}
2	0.02	0.01	9.40×10^{-6}
3	0.04	0.01	1.88×10^{-5}
4	0.02	0.02	1.88×10^{-5}

> **REMEMBER:** When calculating the order with respect to the concentration of a reactant, the concentration of other reactants must remain the same.

Comparing Experiments 1 and 2 in the table above:

if we double $[H_2]$ but keep $[I_2]$ the same, the rate doubles from 4.70×10^{-6} to 9.40×10^{-6}.

So the order of reaction with respect to $[H_2]$ is 1.

Comparing Experiments 2 and 4:

if we double $[I_2]$ but keep $[H_2]$ the same, the rate doubles from 9.4×10^{-6} to 1.88×10^{-5}.

So the order of reaction with respect to $[I_2]$ is 1.

PRACTICE QUESTIONS

1   The table below shows the change in concentration of bromine during the course of a reaction.

Time/s	$[Br_2]/mol\,dm^{-3}$
0	0.0100
60	0.0090
120	0.0066
180	0.0053
240	0.0044
360	0.0028

   a   Plot a concentration–time graph for the data in the table.

   b   Calculate the half-life of Br_2 by drawing tangents.

   c   Find the half-life at two points and deduce the order of the reaction.

2   The table below relates to the reaction between hydrogen and nitrogen oxide.

Experiment	Initial [NO] /$mol\,dm^{-3}$	Initial [H$_2$] /$mol\,dm^{-3}$	Initial rate /$mol\,dm^{-3}\,s^{-1}$
1	6×10^{-3}	1×10^{-3}	3.0×10^{-3}
2	6×10^{-3}	2×10^{-3}	6.0×10^{-3}
3	6×10^{-3}	3×10^{-3}	9.0×10^{-3}
4	1×10^{-3}	6×10^{-3}	0.5×10^{-3}
5	2×10^{-3}	6×10^{-3}	2.0×10^{-3}
6	3×10^{-3}	6×10^{-3}	4.5×10^{-3}

   a   State what happens to the initial rate when [H$_2$] is doubled and [NO] remains the same.

   b   Determine the order with respect to [H$_2$].

   c   State what happens to the initial rate when [NO] is doubled and [H$_2$] remains the same.

   d   Determine the order with respect to [NO].

STRETCH YOURSELF!

What percentage is left?

The half-life of carbon-14 is 5800 years. It accumulates in living plants and decays after death.

   a   What percentage of carbon-14 will remain 16 000 years after the plant has died?

   b   Construct a table and plot a graph of fraction of carbon-14 left vs number of half-lives.

   c   Calculate how long it would take for the concentration of carbon-14 to fall to 40% of the original concentration.

6.4 Rates by inspection 1

Using trends in concentration data to deduce the orders of reactant

The order of reaction can be found by finding the initial rate at the start of the experiment when all of the concentrations are known. You then find the initial rate for different concentrations. This method is known as the initial rate method.

The concentration of a reactant can be measured during the experiment and the rate can be calculated at intervals. This is known as the continuous method. The order can then be deduced from a concentration–time graph.

For example, if rate = $k[A][B]^2$

the total order is the sum of orders: $1 + 2 = 3$

✓ WORKED EXAMPLE

Calculating orders of reaction using the initial rate method

$A + B \rightarrow C$

Experiment	[A]/mol dm^{-3}	[B]/mol dm^{-3}	Rate/mol dm^{-3} s^{-1}
1	0.01	0.01	2.1×10^{-9}
2	0.01	0.02	4.2×10^{-9}
3	0.02	0.02	4.2×10^{-9}

Comparing Experiments 1 and 2:

when [B] = 0.01 the rate = 2.1×10^{-9}

when [B] = 0.02 the rate = 4.2×10^{-9}, which is double the rate when [B] = 0.01

So when the concentration of reactant B doubles so does the rate.

The order of reaction with respect to reactant [B] is 1.

When the concentration of A doubles (Experiment 2 to Experiment 3) the rate does not change.

The order of reaction with respect to reactant [A] is zero.

So the rate equation is:

rate = $k[B]$

The total order of the reaction is 1.

> **REMEMBER:** To find the order of the reaction with respect to a reactant, all other reactants have to be kept at a constant concentration.

? PRACTICE QUESTIONS

1 Use the data in this table to answer the questions that follow:

Experiment	[A]/mol dm^{-3}	[B]/mol dm^{-3}	[C]/mol dm^{-3}	Rate/mol dm^{-3} s^{-1}
1	0.002	0.001	0.001	0.180
2	0.004	0.001	0.001	0.700
3	0.002	0.002	0.001	0.180
4	0.002	0.001	0.002	0.180

a Calculate the order of each reactant.

b Write the rate equation.

c Determine the total order of the reaction.

2 Use the data in this table to answer the questions that follow:

Experiment	$[X]/mol\,dm^{-3}$	$[Y]/mol\,dm^{-3}$	$[Z]/mol\,dm^{-3}$	$Rate/mol\,dm^{-3}\,s^{-1}$
1	0.50	0.20	0.10	0.0078
2	0.20	0.20	0.10	0.0077
3	0.10	0.20	0.10	0.0081
4	0.50	0.40	0.10	0.0312
5	0.50	0.10	0.10	0.0019
6	0.50	0.20	0.05	0.0040
7	0.50	0.20	0.20	0.0157

a Calculate the order of each reactant.

b Write the rate equation.

c Determine the total order of the reaction.

STRETCH YOURSELF!

The rate-determining step

$M + N + P \rightarrow Q$

The equation above is a balanced reaction.

The rate equation is:

$Rate = k[M]^2[N][P]^0 = k[M]^2[N]$

A proposal for the mechanism is:

$2M + N \rightarrow M_2N$ Slow rate-determining step

We can deduce this is the rate-determining step because the concentrations of both these reactants affect the rate.

$M_2N + P \rightarrow Q$ Fast step

We can deduce this is a fast step because the concentration of P does not affect the rate of reaction.

PRACTICE QUESTION

3 The rate equation for the reaction of NO_2 gas with CO gas is:

$Rate = k[NO_2]^2$

The proposed mechanism is:

$2NO_2(g) \rightarrow NO_3(g) + NO(g)$

$NO_3(g) + CO(g) \rightarrow NO_2(g) + CO_2(g)$

$NO(g) + \frac{1}{2}O_2(g) \rightarrow NO_2(g)$

a Give the overall reaction.

b Identify the rate-determining step and explain why.

6.5 Rates by inspection 2

Writing a rate equation based on the rate constant, k, and the orders of the reactants

A rate equation summarises how the concentration of reactants affects the rate of reaction. Consider the chemical equation:

$$xA + yA \rightarrow zAB$$

Rate $= k[A]^n[B]^m$

where n and m are the orders with respect to reactants [A] and [B].

Note: Different letters are used for the orders of reactants (n and m) to the stoichiometric coefficents ($x, y,$ and z) because orders are determined experimentally and are not due to stoichiometry.

WORKED EXAMPLE

How to write the rate equation for the reaction of hydrogen and iodine to form hydrogen iodide

$H_2(g) + I_2(g) \rightarrow 2HI(g)$

From experiment it was found that:

The order of reaction with respect to $[H_2] = 1$

The order of reaction with respect to $[I_2] = 1$

The overall order of the reaction with respect to both reactants is 2.

The rate equation can be written:

Rate $= k[H_2]^1[I_2]^1$

$[H_2]^1 = [H_2]$

More simply, rate $= k[H_2][I_2]$

> **REMEMBER:** A reactant that is zero order does not need to be included in the rate equation, because its concentration does not affect the rate. $[A]^0 = 1$

PRACTICE QUESTIONS

1 $2NO(y) + 2H_2(g) \rightarrow 2H_2O(g) + N_2(g)$

   The reaction is second order with respect to [NO] and first order with respect to $[H_2]$. Write the rate equation.

2 $A + B \rightarrow C + D$

   The reaction is third order with respect to [A] and zero order with respect to [B]. Write the rate equation.

3 Rate $= k[A]^1[B]^2[C]^3$

   Explain what would happen to the initial rate if you doubled the concentration of:

   a A

   b B

   c C

STRETCH YOURSELF!

What are the units of the rate constant?

The units of the rate constant are dependent on the order of the reaction.

Example

Equating units:

$$\text{Rate} = k[A][B] = \text{mol}\,\text{dm}^{-3}\,\text{s}^{-1}$$

$$k = \frac{\text{rate}}{[A][B]} = \frac{\text{mol}\,\text{dm}^{-3}\,\text{s}^{-1}}{\text{mol}\,\text{dm}^{-3} \times \text{mol}\,\text{dm}^{-3}}$$

$$k = \frac{\text{mol}\,\text{dm}^{-3}\,\text{s}^{-1}}{\text{mol}^2\,\text{dm}^{-6}} = \text{mol}^{-1}\,\text{dm}^3\,\text{s}^{-1}$$

Calculate the units of the rate constant for:

$$\text{Rate} = k[C]^3[D]^0$$

PRACTICE QUESTIONS

For the gaseous reaction

$$A(g) + B(g) \rightarrow C(g) + D(g)$$

$$\text{Rate} = k[A]^2[B]$$

Calculate how many times the rate increases if:

4 The concentration of both A and B doubles.

5 The concentration of A doubles but the concentration of B remains constant.

6 The volume of the reaction vessel is doubled.

6.6 Rates (finding the value and units of the rate constant)

Calculating a constant by rearranging an equation

The rate equation is a simple way of writing how the rate of reaction is changed by the concentration of the reactants.

$$Rate = k[A]^m[B]^n$$

- [A] and [B] are the concentrations of the reactants.
- m and n are the orders of the reactants.
- k is the rate constant, which is unique for a particular temperature.
- Rate is measured in $mol\,dm^{-3}\,s^{-1}$.

In order to find the rate constant and the units we have to rearrange the equation to:

$$k = \frac{rate}{[A]^m[B]^n}$$

The units of the rate constant are dependent on the overall order of the reaction ($m + n$).

WORKED EXAMPLE

How to calculate the units of the rate constant

$$Rate = k[A][B]^2$$

$$k = \frac{rate}{[A][B]^2}$$

$$Units\ of\ k = \frac{\cancel{mol\,dm^{-3}}\,s^{-1}}{mol\,dm^{-3} \times mol\,dm^{-3} \times \cancel{mol\,dm^{-3}}}$$

$$Units\ of\ k = mol^{-2}\,dm^6\,s^{-1}$$

How to calculate the rate constant

At 500 K the rate of reaction was found to be $0.00350\,mol\,dm^{-3}\,s^{-1}$

The concentration of A was $0.0015\,mol\,dm^{-3}$.

The concentration of B was $0.0045\,mol\,dm^{-3}$.

$$k = \frac{0.0035}{(0.0015)(0.0045)^2} = 1.15 \times 10^5\,mol^{-2}\,dm^6\,s^{-1}$$

> REMEMBER: Make sure you calculate the units of the rate constant first. This will help avoid careless mistakes after you have come up with the correct rate constant.

PRACTICE QUESTIONS

1 The concentration of H_2 was $0.1\,mol\,dm^{-3}$ and the concentration of I_2 was $0.2\,mol\,dm^{-3}$. They were mixed at 450 °C. The initial rate of reaction was $2.3 \times 10^{-5}\,mol\,dm^{-3}\,s^{-1}$.

$$H_2 + I_2 \rightarrow 2HI \quad rate\ equation = k[H_2][I_2]$$

a What are the units of k?

b Calculate the constant k.

2 The concentration of H_2 was $0.01\,mol\,dm^{-3}$ and the concentration of NO was $0.02\,mol\,dm^{-3}$. The initial rate at $850\,°C$ was $0.00345\,mol\,dm^{-3}\,s^{-1}$.

$$2NO + 2H_2 \rightarrow 2H_2O + N_2 \quad rate = k[NO][H_2]$$

a Give the units of k.

b Calculate the constant k.

3 The temperature of the reaction in Question 2 was kept the same, but the concentration of H_2 was doubled and the rate quadrupled.

a Rearrange the equation to make [NO] the subject.

b Calculate the new concentration of NO.

c Suggest what effect increasing the temperature would have on the rate constant k.

STRETCH YOURSELF!

Rates and the rate-determining step

Many reactions take place in more than one step. However, some steps take place so quickly that they have negligible influence on rate in comparison with slower steps.

Consider three reactants [A], [B], and [C].

The equation is $A + B + C \rightarrow D$

$A + B \rightarrow AB$ is the slow step and so determines the rate.

$AB + C \rightarrow D$ is fast.

The concentrations of A and B and C are varied, and the initial rate of reaction is measured. The value for the first experiment is given in the table.

> **REMEMBER:** If the concentration of a reactant appears in the rate equation it should take part in the slow step. If it does not appear in the rate equation the rate is zero order with respect to that reactant.

Experiment	[A]	[B]	[C]	Rate / $mol\,dm^{-3}\,s^{-1}$
1	0.1	0.2	0.1	0.00308
2	0.2	0.2	0.1	
3	0.3	0.2	0.1	
4	0.1	0.4	0.1	
5	0.1	0.6	0.1	
6	0.1	0.2	0.2	
7	0.1	0.2	0.3	

a Use the mechanism to deduce the order with respect to [A], [B], and [C].

b Write the rate equation for this reaction.

c Calculate the rate constant and give its units.

d Copy the final column of the table and use the rate equation to complete it.

6.7 Finding the activation energy of a reaction

Using a graphical method to find the activation energy of the reaction

To find the activation energy of the reaction through graphical methods, the Arrhenius equation must be used. This will be provided for you in the examination or core practical assessment.

$$k = A^e - \frac{E_a}{RT}$$

k is the rate constant which is proportional to reaction rate. A is a constant related to the number of collisions between reactant molecules. $e^{(-E_a/RT)}$ is the fraction of collisions with enough energy to react.

An experiment is run at different temperatures. The variables are rate constant and temperature, with activation energy remaining constant.

To find the activation energy, E_a, the Arrhenius equation must be expressed as a logarithmic relationship, so it can be plotted as a straight line in the format $y = mx + c$.

$$\ln k = -\frac{E_a}{RT} + \ln A$$

A graph of $\ln k$ against $\frac{1}{T}$ has a gradient of $-\frac{E_a}{R}$

Note: k is the rate constant, but this constant is unusual because it varies with temperature. The rate constant expression is simply the fraction of molecules with the activation energy at a particular temperature.

WORKED EXAMPLE

The decomposition of dinitrogen pentoxide, N_2O_5

Decomposition is carried out between 0 and 65 °C.

REMEMBER: You need to convert your temperature to kelvin for the Arrhenius equation.

You must first carry out the experiment at 0 °C and measure the time for N_2O_5 to decompose. This then needs to be repeated for the other temperatures.

T/K	k/s^{-1}	$\frac{1}{T}/K^{-1}$	$\ln k$
273	0.0787	0.003 66	−2.54211
298	3.46	0.003 36	1.24127
308	13.5	0.003 25	2.60269
318	49.8	0.003 14	3.90801
328	150	0.003 05	5.01064
338	487	0.002 96	6.18826

WORKED EXAMPLE

Finding the activation energy from the gradient

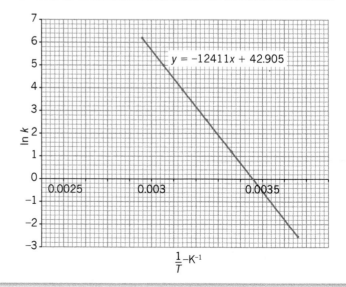

$y = -12411x + 42.905$

The gradient $= -\dfrac{E_a \,(\text{activation energy})}{R \,(\text{gas constant})}$

We can then calculate the activation energy, E_a. ($R = 8.131\,41\,\text{J}\,\text{mol}^{-1}\,\text{K}^{-1}$)

$\dfrac{E_a}{R} = -12411$

$E_a = 12411 \times 8.3141 = 103186\,\text{J}\,\text{mol}^{-1}$

This then needs to be divided by 1000 to convert to $\text{kJ}\,\text{mol}^{-1}$.

$E_a = 103.2\,\text{kJ}\,\text{mol}^{-1}$

PRACTICE QUESTIONS

1 In an experiment, methyl isocyanide is converted to acetonitrile through the process of isomerisation.

The experiment is performed at the following temperatures.

T/K	k/s^{-1}	$\dfrac{1}{T}$	$\ln k$
480	1.3598×10^{-4}	0.002 08	−8.903
490	3.0998×10^{-4}	0.002 04	−8.079
500	6.8097×10^{-4}	0.002 00	−7.292
510	1.4503×10^{-3}	0.001 96	−6.536
520	2.1941×10^{-3}	0.001 92	−6.122

a Plot a graph of $\ln k$ vs $\dfrac{1}{T}$. Draw a straight line through the points.

b Find the gradient of the line, $-\dfrac{E_a}{R}$

c Find the activation energy, E_a of the reaction.

The gas constant R is $8.131\,41\,\text{J}\,\text{mol}^{-1}\,\text{K}^{-1}$

2 An industrial process has been improved using a catalyst. Plot a graph of $\ln k$ against $\dfrac{1}{T}$ for both processes.

Process A

T/K	k/s^{-1}
300	0.0133
310	0.0400
320	0.1111
330	0.3333
340	1.0000

Process B

T/K	k/s^{-1}
300	0.0007
310	0.0022
320	0.0087
330	0.0286
340	0.1000

a Identify from the table which process uses the catalyst.

b Find the gradient of the line for processes A and B.

c Calculate the activation energy of each process.

d Calculate the percentage reduction in activation energy obtained when using the catalyst.

e Suggest reasons why the catalyst may not be used.

SUMMARY QUESTIONS FOR CHAPTERS 4–6

1 One possible method for the production of ethanol is the fermentation of sugar from plants to produce ethanol and carbon dioxide. The reaction is catalysed by the enzymes in yeast.
$C_6H_{12}O_6 \rightarrow 2C_2H_5OH + 2CO_2$
(A_r: C, 12; O, 16; H, 1)

a Calculate the atom economy for this method of producing ethanol. Give you answer to three significant figures.

b If 7.20 g of glucose was added, calculate the maximum yield of ethanol that could be produced in the experiment.

c The experiment is carried out and 2.76 g of ethanol was produced. Calculate the percentage yield of the reaction.

2 Bromine reacts with ethene to produce 1,2-dibromoethane.
The displayed formula for the reaction is shown below.

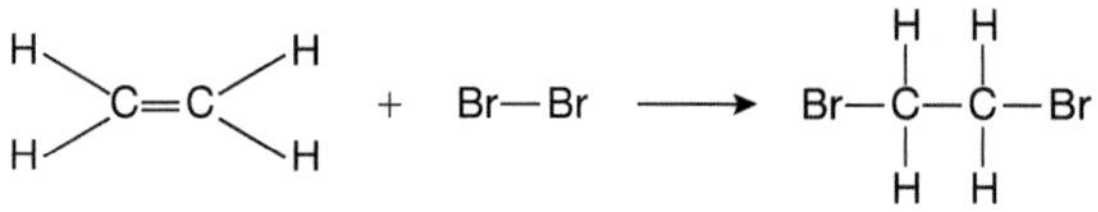

a What is the atom economy for the reaction? Explain why a calculation is not required.

b An experiment is carried out and the enthalpy change for the reaction is found to be −115 kJ mol^{-1}. Use the bond enthalpy values shown in the table below to determine the bond enthalpy value for the C=C bond.

Bond	Bond enthalpy value / kJ mol^{-1}
C–H	+410
Br–Br	+190
C–C	+347
C–Br	+285

3 The mass spectrum for propanal shows a number of peaks.
Identify the ions responsible for the peaks at:
a m/z 15   b m/z 29   c m/z 43   d m/z 58

4 A chemist carries out a titration to find the volume of sodium hydroxide needed to neutralise 25 cm^3 of a solution of 0.15 mol dm^{-3} nitric acid.
The reaction is shown below.
$NaOH(aq) + HNO_3(aq) \rightarrow NaNO_3(aq) + H_2O(l)$
The results for the volume of sodium hydroxide required are shown below. The burette has a maximum error of plus or minus 0.05 cm^3.

Experiment	Volume of sodium hydroxide / cm^3
1	22.95
2	22.60
3	22.30
4	22.70

a Calculate the average volume of nitric acid used.

b Calculate the maximum percentage error in the volume of sodium hydroxide in Experiment 1. Give your answer to three significant figures.

5 A student carried out the iodination of propanone at five different concentrations of bromine water and propanone to find the rate equation.
$CH_3COCH_3 + I_2 \rightarrow CH_3COCH_2I + H^+ + I^-$

94

Rate/s^{-1}	[CH$_3$COCH$_3$]/mol dm^{-3}	[I$_2$]/mol dm^{-3}	[H$^+$]/mol dm^{-3}
5 × 10^{-5}	0.50	0.05	0.05
5 × 10−5	0.50	0.10	0.05
1.5 × 10^{-4}	0.50	0.05	0.10
3.0 × 10^{-4}	0.75	0.05	0.20
9.0 × 10^{-5}	0.75	0.05	0.05

a   Determine the order of the reaction to [I$_2$] and [CH$_3$COCH$_3$], [H$^+$], and determine the overall order of the reaction.

b   Give the rate equation.

c   Calculate the units of the rate constant, k.

d   Describe the effect of increasing the temperature on the rate constant.

The mechanism for the reaction is below:

Step 1 (CH$_3$)$_2$C=O + H$^+$ ⇌ (CH$_3$)$_2$C=O$^+$H

Step 2 (CH$_3$)$_2$C=O + H ⇌ CH$_3$C(OH)=CH$_2$ + H$^+$

Step 3 CH$_3$C(OH)=CH$_2$ + I$_2$ → CH$_3$COCH$_2$I + HI

e   Identify the rate-determining step in the reaction and justify your answer.

6   Radioactive uranium decays to lead. The half-life of uranium is 4.5×10^9 years.

A sample of rock was analysed using this process and it was found to contain 4.0×10^{-3} g of uranium and $3 \times 10 \times 10^{-5}$ g of lead.

$$\ln \frac{N_t}{N_0} = -kt$$

a   How long would it take half the sample of uranium to decay?

b   Plot a graph of fraction of uranium left vs number of half lives.

c   Suggest why some countries are reluctant to store uranium or radioactive waste.

d   Calculate the decay constant using the following values:

t = the half-life of uranium

N_0 = 2 (the proportion of uranium atoms at $t = 0$)

N_t = 1 (the proportion of uranium atoms after one half-life)

e   How old is the sample?

7   According to the Gibbs free energy equation, if the free energy, G, of the system is negative, then the reaction is spontaneous. Calculate the Gibbs free energy of the reactions below, and determine whether these reactions are spontaneous at 298 K.

a   W + X → Y + Z

b   Y + Z → W + X

c   W + Y → X + Z

d   W + Z → X + Y

Species	$\Delta Hf°$/kJ mol^{-1}	$S°$/J K^{-1} mol^{-1}
W	−390.0	180.0
X	−140.0	235.0
Y	−35.0	280.0
Z	−295.0	300.0

8   The proton NMR spectrum is given below. Compound X has the molecular formula C$_4$H$_8$O$_2$.

It forms an orange precipitate with Brady's reagent (2,4 DNPH) but does not react with Tollen's reagent. Identify compound X.

δ/ppm	2.20	2.69	3.40	3.84
Splitting pattern	singlet	triplet	singlet	triplet
Integration value	3	2	1	2

7.1 Equilibrium equations (writing equations and finding the units)

Calculating the value of the equilibrium constant including its data from units

The Haber process is an important industrial process that produces ammonia in equilibrium.

$$N_2(g) + 3H_2(g) \rightleftharpoons 2NH_3(g)$$

An industrial chemist needs to be able to calculate the amount of product produced at equilibrium.

The law of equilibrium is a quantitative law that requires mathematical proficiency in rearranging calculations in order to predict the amount of product and reactants when a reversible reaction reaches dynamic equilibrium.

The amount of reactants and products formed at equilibrium can be predicted at a certain temperature. To predict the equilibrium concentrations, the equilibrium constant, K_c, must be found.

The general equation for a reversible reaction is:

$$aA + bB \rightleftharpoons cC + dD$$

The expression for the equilibrium constant is:

$$K_c = \frac{[C]^c [D]^d}{[A]^a [B]^b}$$

> **REMEMBER: Don't** get confused! A lower case 'k' is rate constant and an upper case 'K' is equilibrium constant

The letters in brackets, for example, [C], relate to the concentration of reactants or products in $mol\,dm^{-3}$.

The powers, for example, $[\ ]^c$, relate to the stoichiometric number of molecules in the equation.

WORKED EXAMPLE

The Haber process

The equilibrium constant for the Haber process is calculated using the following equation:

$$K_c = \frac{[NH_3]^2}{[N_2][H_2]^3}$$

To work out the units of the equilibrium constant:

$$K_c = \frac{mol\,dm^{-3} \times mol\,dm^{-3}}{mol\,dm^{-3} \times mol\,dm^{-3} \times mol\,dm^{-3} \times mol\,dm^{-3}}$$

$$K_c = \frac{1}{mol\,dm^{-3} \times mol\,dm^{-3}}$$

$$K_c = \frac{1}{mol^2\,dm^{-6}} \text{ or } mol^{-2}\,dm^{6}$$

> **REMEMBER: The units of** the equilibrium constant, K_c, are determined by the stoichiometric ratio of reactants and products. K_c itself is determined by experiment.

How to calculate an equilibrium constant

A mixture reached equilibrium in a $0.25\,dm^3$ reaction vessel and consisted of $7.4\,mol$ of nitrogen, $1.6\,mol$ hydrogen, and $0.40\,mol$ of ammonia.

Concentration is measured in $mol\,dm^{-3}$.

To calculate the concentration of the reactants and products:

$$Concentration = \frac{no.\ of\ moles}{volume\ in\ dm^3}$$

$$[N_2] = \frac{7.4}{0.25} = 29.6\,mol\,dm^{-3}$$

$$[H_2] = \frac{1.6}{0.25} = 6.4\,mol\,dm^{-3}$$

$$[NH_3] = \frac{0.4}{0.25} = 1.6 \, mol \, dm^{-3}$$

$$K_c = \frac{(1.6)^2}{(29.6)(6.4)^3} = 3.3 \times 10^{-4} \, mol^{-2} dm^6$$

PRACTICE QUESTIONS

1 This question is about the contact process. In the second stage of the contact process, sulfur dioxide is oxidised to sulfur trioxide. The forward reaction is exothermic.

$$2SO_2(g) + O_2(g) \rightleftharpoons 2SO_3(g)$$

  a Give the expression for the equilibrium constant, K_c.

  b Calculate the units of K_c.

  c The mixture is kept in a $10\,000 \, dm^3$ vessel. At equilibrium there are 60 moles of SO_2, 30 moles of O_2, and 1920 moles of sulfur trioxide. Calculate K_c.

  d Predict what would happen to the yield of sulfur trioxide if the temperature of the equilibrium was increased.

2 This question is about the decomposition of dinitrogen tetroxide, N_2O_4. This decomposes on heating to form NO_2. The reactions exist in a dynamic equilibrium:

$$N_2O_4(g) \rightleftharpoons 2NO_2(g)$$

  a Give the expression for the equilibrium constant, K_c.

  b Calculate the units of K_c.

  c The mixture decomposes in a $1 \, dm^3$ reaction vessel to form an equilibrium mixture containing 5 mol of both gases. Calculate K_c.

  d The reaction is an example of a thermal decomposition reaction. Describe the relationship between temperature and K_c.

3 This question is about the formation of the ester methyl ethanoate, which is formed in equilibrium:

$$CH_3OH + CH_3COOH \rightleftharpoons CH_3COOCH_3 + H_2O$$

6.67 moles of CH_3OH was mixed with 5 moles of CH_3COOH. The mixture was allowed to reach equilibrium, at which point there were 0.67 moles of CH_3COOH. The total volume of solution was $2000 \, cm^3$.

  a Give the expression for the equilibrium constant K_c.

  b Calculate the units of K_c.

  c Determine how many moles of CH_3COOH have reacted.

  d Using the stoichiometric relationship in the equation determine the number of moles of water formed.

  e Determine the equilibrium concentrations of all the reactants and products.

  f Find K_c.

STRETCH YOURSELF!

Using K_c to predict changes in concentration

In the gas phase, hydrogen and iodine react to produce an equilibrium mixture containing hydrogen iodide:

$$H_2(g) + I_2(g) \rightleftharpoons 2HI(g)$$

The equilibrium constant, K_c, for this reaction is 36.0 at 500 K. If the concentration of hydrogen gas is trebled, calculate the change in the hydrogen iodide concentration.

$$K_c = \frac{[HI]^2}{[H_2][I_2]} \qquad \text{Note: In this case there are no units for } K_c.$$

K_c remains constant if there is no increase in temperature.

$$\frac{[HI]^2}{[H_2][I_2]} = 36$$

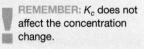

REMEMBER: K_c does not affect the concentration change.

$[HI]^2 = 36[H_2][I_2]$

If $[H_2]$ increases three-fold, then $[HI]^2$ must increase three-fold.

So $[HI] = \sqrt{3} = 1.73$

If $[H_2]$ triples, then $[HI]$ will increase by a factor of 1.73.

7.2 Equilibrium calculations given equilibrium concentrations or partial pressure

Making predictions on concentration of reactants using K_c

In order to calculate the equilibrium constant, you need to know the molar concentrations of the reactants and products.

For a reaction:

$$aA + bB \rightleftharpoons cC + dD$$

$$K_c = \frac{[C]^c[D]^d}{[A]^a[B]^b}$$

K_c can be calculated in two steps.

Step 1: Find the units of K_c.

Step 2: Input molar concentrations into the equation to find K_c.

WORKED EXAMPLE

How to calculate an equilibrium constant given equilibrium concentrations

The hydrolysis of ethyl ethanoate forms an equilibrium at 350 K.

$$CH_3COOCH_3 \rightleftharpoons CH_3COOH + CH_3OH + H_2O$$

$[CH_3COOCH_3] = 0.010 \, mol \, dm^{-3}$

$[CH_3COOH] = 0.0037 \, mol \, dm^{-3}$

$[CH_3OH] = 0.0037 \, mol \, dm^{-3}$

Step 1: $K_c = \dfrac{mol \, dm^{-3} \times \cancel{mol \, dm^{-3}}}{\cancel{mol \, dm^{-3}}} = mol \, dm^{-3}$

Step 2: $K_c = \dfrac{(0.0037)(0.0037)}{(0.010)} = 0.001\,369 \, mol \, dm^{-3} = 1.369 \times 10^{-3} \, mol \, dm^{-3}$

> **REMEMBER:** If you know the molar concentrations of the reactants and products, finding the value for K_c is easy. You need to be careful with the units as these vary and should be worked out first.

The concentration of H_2O is very nearly constant so doesn't appear in the calculation.

PRACTICE QUESTIONS

1 $HOOCCH_2NH_3^+(aq) \rightleftharpoons HOOCCH_2NH_2(aq) + H^+(aq)$

The concentrations at equilibrium are:

$[HOOCCH_2NH_3^+] = 0.04 \, mol \, dm^{-3}$

$[HOOCCH_2NH_2] = 0.20 \, mol \, dm^{-3}$

$[H^+] = 0.20 \, mol \, dm^{-3}$

a Give the expression for K_c.

b Calculate the units of K_c.

c Calculate the value of K_c.

2 The equilibrium of ozone and oxygen was studied in a $2 \, dm^3$ container.

$$2O_3(g) \rightleftharpoons 3O_2(g)$$

The concentrations at equilibrium are:

$[O_3(g)] = 1.4 \, mol \, dm^{-3}$

$[O_2(g)] = 1.4 \, mol \, dm^{-3}$

a Give the expression for K_c.

b Calculate the units of K_c.

c Determine the value of K_c.

d Determine the number of moles at equilibrium.

e Calculate the mass of ozone, O_3, at equilibrium.

Making predictions on partial pressure of reactants using K_p

In a gas phase reaction:

$$aA(g) + bB(g) \rightleftharpoons cC(g) + dD(g)$$

$$K_p = \frac{[P_c]^c [P_d]^d}{[P_A]^a [P_B]^b}$$

> **REMEMBER:** All values for pressures should be in the same units!

You may be given partial pressure values of individual gases instead of concentrations. These can be treated similary to K_c calculations. If the equilibrium is heterogenous (meaning in different states), pure solids and liquids should be given the partial pressure value $P = 1$, as their pressure is constant. Solids and pure liquids, therefore, do not appear in the equilibrium expression.

WORKED EXAMPLE

$2H_2(g) + O_2(g) \rightarrow 2H_2O(l)$

$PH_2(g) = 100\,kPa \qquad PO_2(g) = 110\,kPa \quad P\,H_2O(l) = 1\,kPa$

$K_p = P^2\,H_2O\,/\,P^2\,H_2\,\,P\,O_2$

$\dfrac{1}{(100)^2\,(110)} = 9.09 \times 10^{-7}$

PRACTICE QUESTION

3 The reaction below is a gas phase reaction.

$2O_3(g) \rightleftharpoons 3O_2(g)$

Calculate K_p if the partial pressures at equilibrium are:

$O_3 = 140\,kPa$

$O_2 = 140\,kPa$

a Write the expression for K_p.

b Calculate the units of K_p.

c Find the value of K_p.

STRETCH YOURSELF!

Calculating the concentrations in an equilibrium

The Haber process makes ammonia. Nitrogen and hydrogen are reacted over an iron catalyst:

$N_2(g) + 3H_2(g) \rightleftharpoons 2NH_3(g)$

28 tonnes ($28 \times 1000\,kg$) of nitrogen was reacted with 2 tonnes of hydrogen. The mixture was left to go to equilibrium and 9 tonnes of NH_3 was made. The total volume was $5\,dm^3$.

a Calculate the number of moles of N_2 and H_2 at the start of the reaction.

b Determine how many moles of NH_3 were present at equilibrium.

c Determine how many moles of N_2 were used up.

d Calculate how many moles of H_2 were used up.

e Calculate the number of moles of N_2 and H_2 at equilibrium.

f Calculate the concentrations at equilibrium.

$CH_3OH + CH_3CH_2COOH \rightleftharpoons CH_3COOCH_3 + H_2O$

Methanol and ethanoic acid react to form methyl ethanoate in equilibrium. 5 mol of methanol, 6 mol of ethanoic acid, 6 mol of methyl ethanoate, and 4 mol of water were mixed together in a sealed container at room temperature. At equilibrium there were 4 mol of ethanoic acid in the mixture.

g Give the expression for the equilibrium constant for this reaction.

h Determine how many moles of reactants and products are present at equilibrium.

i Calculate K_c for this reaction.

j Suppose 1 mol of methanol, 1 mol of ethanoic acid, 3 mol of methyl ethanoate, and 3 mol of water are mixed. Determine the concentration of all reactants and products at equilibrium.

7.3 Equilibrium calculations given moles at equilibrium

Showing step-by-step deductions in finding the equilibrium constant, by finding concentrations at equilibrium

So far the worked examples given have shown you how to work out the value of K_c from equilibrium concentrations. However, in some questions the amounts are given in moles and the total volume is also stated. In these problems it is necessary to work out the equilibrium concentrations first in $\mathrm{mol\,dm^{-3}}$.

Use: $\text{concentration} = \dfrac{\text{mol}}{\text{vol (dm}^3)}$ or $\text{concentration} = \dfrac{\text{mol}}{\text{vol (cm}^3)} \times 1000$

WORKED EXAMPLE

How to calculate an equilibrium constant from the number of moles at equilibrium

An equilibrium mixture at 700 K in a 0.25 dm^3 reaction vessel consisted of 3.4 mol of nitrogen, 0.25 mol of hydrogen, and 0.40 mol of ammonia.

Step 1: Calculate the concentrations.

Concentration of N_2 $= \dfrac{3.4}{0.25} = 13.6\,\mathrm{mol\,dm^{-3}}$

Concentration of H_2 $= \dfrac{0.25}{0.25} = 1.0\,\mathrm{mol\,dm^{-3}}$

Concentration of NH_3 $= \dfrac{0.4}{0.25} = 1.6\,\mathrm{mol\,dm^{-3}}$

Step 2: Calculate K_c.

$K_c = \dfrac{(1.6)^2}{(13.6)(1)^3} = 1.88 \times 10^{-1}\,\mathrm{mol^{-2}\,dm^6}$

> **REMEMBER:** The magnitude of K_c can be used to estimate the position of equilibrium. If K_c is less than 1, the position of equilibrium lies towards the left-hand side. If the value of K_c is greater than 1, the position of equilibrium lies to the right-hand side.

WORKED EXAMPLE

When K_c has no units

This example is about the formation of the ester methyl ethanoate, which is formed in equilibrium:

$$CH_3OH + CH_3COOH \rightleftharpoons CH_3COOCH_3 + H_2O$$

At equilibrium there are 0.05 moles of CH_3OH, 0.05 moles of CH_3COOH, 0.07 moles of CH_3COOCH_3, and 0.7 moles of H_2O.

Step 1: Determine the units of K_c.

$K_c = \dfrac{[CH_3COOCH_3][H_2O]}{[CH_3OH][CH_3COOH]}$

Units of $K_c = \dfrac{\cancel{mol\,dm^{-3}} \times \cancel{mol\,dm^{-3}}}{\cancel{mol\,dm^{-3}} \times \cancel{mol\,dm^{-3}}}$ so K_c has no units.

Step 2: Calculate K_c.

$K_c = \dfrac{(0.7)(0.7)}{(0.05)(0.05)} = 196$

PRACTICE QUESTIONS

1 0.1 mol of PCl_3 and 0.1 mol Cl_2 were heated in a 2 dm³ flask at constant temperature. When equilibrium had been established there were 0.05 moles of all reactants and products:

$$PCl_3(g) + Cl_2(g) \rightleftharpoons PCl_5(g)$$

   a Give the expression for K_c for this reaction.

   b Determine the units of K_c.

   c Calculate the value of K_c for this reaction.

2 In the production of ammonia at 800 K, 1.5 mol of NH_3 was produced at equilibrium leaving 2 moles of hydrogen and 0.75 moles of nitrogen in a 3 dm³ reaction vessel.

$$N_2(g) + 3H_2(g) \rightleftharpoons 2NH_3(g)$$

   a Give the expression for K_c for this reaction.

   b Calculate the units of K_c.

   c Calculate the concentrations at equilibrium.

   d Determine the equilibrium constant.

3 In the decomposition of dinitrogen tetroxide the moles at equilibrium are 0.40 moles of NO_2 and 0.50 moles of N_2O_4 in a 0.25 dm³ reaction vessel.

$$N_2O_4(g) \rightleftharpoons 2NO_2(g)$$

   a Give the expression for K_c for this reaction.

   b Calculate the units of K_c.

   c Calculate the concentrations at equilibrium.

   d Determine the equilibrium constant K_c.

STRETCH YOURSELF!

Is it in equilibrium?

0.1 mol each of $SO_3(g)$, $SO_2(g)$, and $O_2(g)$ are introduced into a 1 dm³ container at a temperature at which the equilibrium constant is:
$K_c = 2.0 \times 10^{-6} \, mol^{-1} \, dm^3$.

$$2SO_2(g) + O_2(g) \rightleftharpoons 2SO_3(g)$$

a Show that the mixture is not in equilibrium.

b Predict the concentrations of reactants and product when in equilibrium.

   The following reaction is at equilibrium:

$$CO(g) + 2H_2(g) \rightleftharpoons CH_3OH(g), \quad \Delta H = -92 \, kJ \, mol^{-1}$$

The concentration of reactants and products are 5 mol and 10 mol CO and H_2, respectively. The concentration of CH_3OH is 2 mol.

c When the temperature is lowered the concentrations remain constant. Explain why the system is no longer in equilibrium.

d Predict what effect adding a catalyst to the reaction would have.

e The system restores to equilibrium. At a lower temperature predict what you expect to happen to the concentration of reactants and products.

f In the production of CH_3OH the temperature of the equilibrium will have to be set at a value that is a compromise. Explain why.

7.4 Equilibrium calculations given moles at the start of the reaction

Using a balanced equation to find concentrations of solutions at equilibrium

To find the concentration of reactants and products at equilibrium from initial amounts, you must know the balanced equation for the reaction, the equilibrium expression, the value for K_c, and the direction the reaction will proceed in to reach equilibrium.

These problems require knowledge of skills covered earlier in the book, which should be attempted first.

Step 1: Write a balanced equation for the reaction.

Step 2: Write the ratio of reactants to products, for example, 1 mole CH_4 to 1 mole H_2O.

Step 3: Work out the moles of reactants and products at equilibrium.

Step 4: Write the K_c expression and calculate the units.

Step 5: Calculate the K_c of the equilibrium.

WORKED EXAMPLE

Calculation of the concentration of ethanoic acid and ethyl ethanoate at equilibrium from initial amounts

12 moles of ethanoic acid was added to 25 moles of ethanol. After the system had reached equilibrium only 3 moles of ethanoic acid remained. Calculate K_c for the reaction.

$$CH_3COOH + CH_3CH_2OH \rightleftharpoons CH_3COOCH_2CH_3 + H_2O$$

The reaction is 1 mol : 1 mol : 1 mol : 1 mol.

3 moles of CH_3COOH remained, which means that 9 moles reacted.

Therefore, 9 moles of all reactants reacted and 9 moles of all products were produced.

$25 - 9 = 16$ mol of CH_3CH_2OH remains.

9 moles of $CH_3COOCH_2CH_3$ and 9 moles of H_2O have been produced.

$$K_c = \frac{\cancel{mol dm^{-3}} \times \cancel{mol dm^{-3}}}{\cancel{mol dm^{-3}} \times \cancel{mol dm^{-3}}} \quad \text{so } K_c \text{ has no units.}$$

Volume does not affect K_c.

$$K_c = \frac{(9)(9)}{(3)(16)} = 1.688$$

> **REMEMBER:** It is common to make a mistake when calculating moles of product formed. Writing the ratio is a good check before progressing further.

PRACTICE QUESTIONS

1 In the Haber process, 30 moles of hydrogen are reacted with 15 moles of nitrogen and only 3 moles of nitrogen react.

$$N_2 + 3H_2 \rightleftharpoons 2NH_3$$

a State the ratio of reactants to products.

b Determine the moles of reactants and products at equilibrium.

c Give the K_c expression and state its units.

d Calculate K_c.

2   5 g of blue hydrated copper sulfate is heated leaving 1.4 g of anhydrous copper sulfate.

$$CuSO_4.5H_2O \rightleftharpoons CuSO_4 + 5H_2O$$

  a   Calculate the moles of reactants and products reacted.

  b   Calculate the moles of reactants and products at equilibrium.

  c   Determine the K_c expression and state its units.

  d   Calculate K_c.

3   12.0 g of propanol reacts with 11.5 g of methanoic acid and the mixture is left to form an equilibrium. 75% of the methanoic acid reacts to form propyl methanoate and water.
The total volume of solution is 25.0 cm³.

$$CH_3CH_2CH_2OH + HCOOH \rightleftharpoons HCOOCH_2CH_2CH_3 + H_2O$$

  a   Calculate the concentration of each reactant at the start.

  b   Calculate the moles of reactants and products at equilibrium.

  c   Calculate the concentration of each reactant and product at equilibrium.

  d   Calculate K_c.

STRETCH YOURSELF!

Finding the number of moles required to form a concentration at equilibrium

$$CH_3COOH + CH_3CH_2OH \rightleftharpoons CH_3COOCH_2CH_3 + H_2O$$

Remember the stoichiometric relationship between reactants: 1 mol of CH_3COOH requires 1 mol of CH_3CH_2OH.

  a   Calculate how many moles of ethanol must be mixed with 3 mol of ethanoic acid at 373 K to produce 2 mol of ethyl ethanoate if K_c at the temperature of equilibrium is 14.

  b   How many moles of ethanol must be present at equilibrium at 400 K to produce 2 moles of ethyl ethanoate and water if K_c at the temperature of equilibrium is 16? Give your answer to two significant figures.

  c   Predict what would happen to the equilibrium if one of the reactants was in excess.

  d   4 moles of ethanol was mixed with a solution containing 5 moles of ethanoic acid, 6 moles of ethyl ethanoate, and 3 moles of water to form an equilibrium where only 2 moles of ethanoic acid remained. Determine how many moles of ethanol, ethyl ethanoate, and water must be present at equilibrium.

  e   Calculate the new value for K_c.

7.5 Calculating pH

pH = −log[H⁺]

pH is a logarithmic function of hydrogen ion concentration.

The Brønsted–Lowry definition of an acid is that of a proton donor. You measure the strength of an acid by the concentration of hydrogen ions in solution.

Solutions can range in pH value from 0 to 14. If you consider this in hydrogen ion concentration, the benefits of pH scale are obvious

pH	0	1	7	13	14
H^+(aq)/mol dm^{-3}	1	10^{-1}	10^{-7}	10^{-13}	10^{-14}

A low pH means a high hydrogen ion concentration.

A high pH means a low hydrogen ion concentration.

By using this logarithmic scale you reduce the range of numbers being dealt with to a simple scale of 0–14.

WORKED EXAMPLE

Calculating the pH of strong acids

Nitric acid is a strong acid. Its formula is HNO_3(aq).

Calculate the pH of a solution of nitric acid with a concentration of 0.04 mol dm^{-3}.

For a 0.04 mol dm^{-3} solution of HNO_3(aq) you have a concentration of 0.04 mol dm^{-3} of H^+ ions.

HNO_3(aq) $\rightarrow$ H^+(aq) + NO_3^-(aq)

pH = −log[H^+] = −log[0.04] = 1.4 (1 d.p.)

If only the mass of solute and volume of solvent of the strong acid are given

0.14 g of HCl is dissolved completely in 25 cm^3 of water. Calculate the pH of the solution.

Step 1: Calculate the concentration of the solution in mol dm^{-3}.

Molar mass, M, of HCl = 1 + 35.5 = 36.5 g mol^{-1}

$$\text{No. of moles} = \frac{\text{mass (g)}}{M} \qquad n = \frac{m}{M}$$

No. of moles, $n = \dfrac{0.14}{36.5} = 0.00384$ in 25 cm^3 of water

Concentration $= \left(\dfrac{n}{\text{volume (cm}^3)} \right) \times 1000 = 0.1534$ mol dm^{-3}

Step 2: Calculate the pH of the acid.

pH = −log(0.1534)

pH = 0.8 (1 d.p.)

> **REMEMBER:** Rounding should only be done for answers and not in your working to avoid differences from the exam mark scheme.

PRACTICE QUESTIONS

1 Calculate the pH of solutions with the following H^+ ion concentrations.
Give your answers to one decimal place.

    a $3.77 \times 10^{-3}\,mol\,dm^{-3}$ b $4.86 \times 10^{-4}\,mol\,dm^{-3}$ c $7.00\,mol\,dm^{-3}$

2   0.0007 g of HCl is dissolved completely in $25\,cm^3$ of water. Calculate the pH to one decimal place.

WORKED EXAMPLE
Finding hydrogen ion concentration from pH

It is also important to be able to convert pH back into H^+ ion concentration.
To convert pH into H^+ ion concentration you must put the $[H^+]$ values into standard form.

pH = 2.0 so log $[H^+]$ = −2.0

$[H^+(aq)]\,mol\,dm^{-3} = 10^{-2.0} = 0.01\,mol\,dm^{-3}$

> **REMEMBER:** When comparing the concentration of strong acids that are pH 2.0 and 3.0, there is a factor of 10 between them. If a strong acid has a pH of around 2, you would expect it to have a hydrogen ion concentration of around $0.01\,mol\,dm^{-3}$. Whereas a strong acid with a pH of 3.0 would have a hydrogen ion concentration of around $0.001\,mol\,dm^{-3}$. Knowing this is useful when checking your calculations.

> **REMEMBER:** Working with hydrogen ion concentrations in standard form is easier with a scientific calculator.
> On most calculators it is usually the shift function on the log button. You then enter the −pH value.

PRACTICE QUESTION

3   Calculate the hydrogen ion concentration of solutions with the following pHs.

    a   pH = 4.0   b   pH = 7.0   c   pH = 5.8   d   pH = 9.2

STRETCH YOURSELF!
Diprotic acids

All the acids you have looked at on this page are examples of monoprotic acids, for example, HCl, HNO_3.

The acid sulfuric acid (H_2SO_4) is a diprotic acid. A diprotic acid is an acid that can dissociate twice.

$H_2SO_4(aq) \rightarrow H^+(aq) + HSO_4^-(aq)$

$HSO_4^-(aq) \rightarrow H^+(aq) + SO_4^{2-}(aq)$

PRACTICE QUESTION

4   Investigate the diprotic acid H_2SO_4.

    a   Determine if both dissociations are the same.

    b   Research what the pH of $0.0002\,mol\,dm^{-3}$ $H_2SO_4(aq)$ is and compare it with other strong monoprotic acids.

7.6 The acid dissociation constant

Deducing expressions for K_a and pK_a of weak acids

The Lowry–Brønsted definition of an acid is that of a proton donor. In the last topic, the pH values of strong acids were calculated. A strong acid completely dissociates in solution, donating all its protons. This makes it fairly easy to deduce the concentration of hydrogen ions as it is the same as the concentration of the acid.

A weak acid only partially dissociates in solution, leaving an equilibrium. In order to determine the strength of the acid, you need to know the degree to which it dissociates into H^+ and A^- ions.

$$HA(aq) + H_2O(l) \rightleftharpoons H_3O^+(aq) + A^-(aq)$$

Or expressed more simply:

$$HA(aq) \rightleftharpoons H^+(aq) + A^-(aq)$$

Expressing this as an equilibrium constant in terms of concentrations:

$$K_c = \frac{[H^+(aq)][A^-(aq)]}{[HA(aq)]}$$

The equilibrium constant in this case is known as the acid dissociation constant K_a.

WORKED EXAMPLE

Writing expressions of K_a for specific weak acids
Methanoic acid, HCOOH

Methanoic acid is a weak acid that occurs naturally in ant bites.

$$HCOOH(aq) \rightleftharpoons H^+(aq) + HCOO^-(aq)$$

The K_a of methanoic acid can be expressed as:

$$K_{a\,HCOOH} = \frac{[H^+][HCOO^-]}{[HCOOH]}$$

The acid dissociation constant of methanoic acid is $1.6 \times 10^{-4}\,mol\,dm^{-3}$.

> **REMEMBER:**
> Unlike other equilibrium constants, the units of K_a are always the same: $mol\,dm^{-3}$.
>
>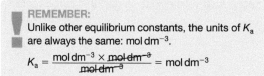
> $$K_a = \frac{mol\,dm^{-3} \times \cancel{mol\,dm^{-3}}}{\cancel{mol\,dm^{-3}}} = mol\,dm^{-3}$$

Acid	$K_a / mol\,dm^{-3}$
methanoic acid	1.6×10^{-4}
benzoic acid	6.3×10^{-5}
3-chlorobutanoic acid	1.0×10^{-4}

Using pK_a to express the dissociation of a weak acid

Just as hydrogen ion concentration, $[H^+]$, can be expressed as a logarithmic function, pH, so can the acid dissociation constant, K_a, as pK_a. This is useful because pK_a has a smaller range of values and is easier to find than the pH of the solution.

Finding pK_a from K_a:

$$pK_a = -\log K_a$$

Finding K_a from pK_a:

$$K_a = 10^{-pK_a}$$

WORKED EXAMPLE

Calculating the pK_a of a weak acid

The K_a of a weak acid is 3.4×10^{-2} mol dm^{-3}.

$pK_a = -\log(3.4 \times 10^{-2}) = 1.5$ (1 d.p.)

To find the K_a value when given the pK_a. The pK_a of a weak acid is 1.5.

$K_a = 10^{-1.5} = 3.2 \times 10^{-2}$ mol dm^{-3} (1 d.p.)

> **REMEMBER:**
> The higher the degree of dissociation is, the larger the acid dissociation constant K_a but the smaller the pK_a value. This is identical to the relationship between $[H^+]$ and pH.

PRACTICE QUESTIONS

1   Give the expression for acid dissociation constant K_a for the following:

   a   ethanoic acid, CH_3COOH

   b   benzoic acid, C_6H_5COOH

   c   3-chlorobutanoic acid, $CH_3CHClCH_2COOH$.

2   Using the table on the previous page, calculate the pK_a values to one decimal place of:

   a   methanoic acid

   b   benzoic acid

   c   3-chlorobutanoic acid.

STRETCH YOURSELF!

Hydrofluoric acid – a weak acid?

Hydrofluoric acid is classed as a weak acid because of its low degree of dissociation into ions.

The acid dissociation constant, K_a, is 7.4×10^{-4} mol dm^{-3}.

Despite being a weak acid, hydrofluoric acid is extremely corrosive and toxic. It also cannot be stored in glass as it reacts with the glass. Instead, it must be kept in plastic polytetrafluoroethylene (PTFE) containers.

a   Give the acid dissociation constant expression for hydrofluoric acid, HF.

b   Calculate the pK_a value.

c   Research why hydrofluoric acid is considered dangerous despite being a weak acid.

7.7 The ionisation of water

Calculating the ionic product of water

Water is a solvent that can function as both an acid or a base, accepting or donating protons.

$$H^+(aq) + H_2O(l) \rightarrow H_3O^+(aq)$$

Here, water is acting as a base, accepting protons.

$$H_2O(l) \rightarrow H^+(aq) + OH^-(aq)$$

Here, water is acting as an acid, donating protons. Water therefore exists in equilibrium between acid and base.

$$2H_2O(l) \rightleftharpoons H_3O^+(aq) + OH^-(aq)$$

Or more simply:

$$H_2O(l) \rightleftharpoons H^+(aq) + OH^-(aq)$$

Water is therefore slightly ionised and the equilibrium constant for this ionisation is given by:

$$K_c = \frac{[H^+]\,[OH^-]}{[H_2O]}$$

Due to the small proportion of water molecules that ionise you can assume that the number of water molecules is a constant.

K_c in this case is known as the ionic product of water, K_w, and the equilibrium equation can be simplified to:

$$K_w = [H^+(aq)][OH^-(aq)]$$

At 298 K, the ionic product of water, K_w, is $10^{-14}\,mol^2\,dm^{-6}$.

✓ WORKED EXAMPLE

Problems involving the ionic product of water

1 Use the ionic product of water to calculate the concentration of both H^+ ions and OH^- ions in water at 298 K.

The pH of water is 7, so the concentration of both ions must be equal.

$$[H^+(aq)] = [OH^-(aq)]$$

At 298 K, the ionic product of water, K_w, is $10^{-14}\,mol^2\,dm^{-6}$.

$$K_w = [H^+(aq)][OH^-(aq)]$$

$$10^{-14} = [H^+(aq)]^2$$

$$[H^+(aq)] = 10^{-7}\,mol\,dm^{-3}$$

$$[OH^-(aq)] = 10^{-7}\,mol\,dm^{-3}$$

2 Calculate the concentration of OH^- ions in a solution of $0.002\,mol\,dm^{-3}$ $HNO_3(aq)$ at 298 K.

At 298 K, the ionic product of water, K_w, is $10^{-14}\,mol^2\,dm^{-6}$.

$$K_w = [H^+(aq)][OH^-(aq)]$$

Rearranging the equation to make OH^- the subject of the equation:

$$[OH^-(aq)] = \frac{K_w}{[H^+(aq)]}$$

$$[OH^-(aq)] = \frac{10^{-14}}{2 \times 10^{-3}} = 5 \times 10^{-12}\,mol\,dm^{-3}$$

> **REMEMBER:**
> Concentrations should be converted to standard form before calculating. This makes the sum easier to perform on a calculator.
>
> When multiplying numbers that are in standard form, remember to add the powers.
>
> When dividing numbers that are in standard form, remember to subtract the powers. This is only possible if the numbers have the same base. In equilibrium calculations, all numbers are to base 10.

PRACTICE QUESTIONS

1 Calculate the concentration of OH^- ions in the following solutions at 298 K:

   a $0.1\,mol\,dm^{-3}$ NaOH

   b $0.04\,mol\,dm^{-3}$ KOH

   c $0.0001\,mol\,dm^{-3}$ HCl

   d $0.0005\,mol\,dm^{-3}$ HNO_3

2 The ionic product of water is only $10^{-14}\,mol\,dm^{-3}$ at 298 K. Calculate the ionic product of water at the following unknown temperatures:

   a $[H^+] = 10^{-6}\,mol\,dm^{-3}$ $[OH^-] = 10^{-6}\,mol\,dm^{-3}$

   b $[H^+] = 10^{-11}\,mol\,dm^{-3}$ $[OH^-] = 10^{-11}\,mol\,dm^{-3}$

   c $[H^+] = 0.000\,02\,mol\,dm^{-3}$ $[OH^-] = 0.000\,02\,mol\,dm^{-3}$

STRETCH YOURSELF!

How does the ionisation of water change with temperature?

The ionic product of water, K_w, is a constant $10^{-14}\,mol\,dm^{-3}$ at 298 K.

Find out how this changes with temperature and what effect, if any, this has on pH.

PRACTICE QUESTIONS

3 $H_2O \rightleftharpoons H^+ + OH^-$ $\Delta H = +58\,kJ\,mol^{-1}$

   a Explain why $[H^+] = [OH^-]$ in pure water.

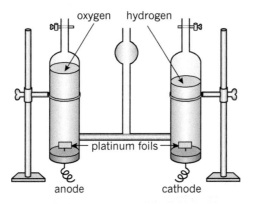

   b A diagram of a Hofmann voltameter is shown above. It is an electrolytic cell used to demonstrate the electrolysis of water. Explain why the water needs to be acidified for electrolysis to work.

   c Explain the effect of temperature on K_w using Le Chatelier's principle.

7.8 pH of strong bases

Using the ionic product of water, K_w, to calculate the pH of a strong base

Strong bases such as sodium hydroxide, $NaOH(aq)$, fully dissociate in solution to hydroxide ions. These react with hydrogen ions to form water. By knowing the concentration of hydroxide ions in solution and the ionic product of water, it is possible to calculate the concentration of hydrogen ions left in solution and therefore the pH.

The pH of strong bases links:

- the ionic product of water $K_w = [H^+(aq)][OH^-(aq)]$
- $pH = -\log[H^+(aq)]$

Refer to topics 7.5 and 7.7 for more detail.

WORKED EXAMPLE

Calculating the pH of strong bases

Sodium hydroxide is a strong base. Calculate the pH of 0.001 mol dm^{-3} solution.

Method 1

Step 1: Calculate the H$^+$ ion concentration in solution.

At 298 K, the ionic product of water, K_w, is 10^{-14} mol^2 dm^{-6}.

Concentration of OH$^-$ = 0.001 mol dm^{-3} = 10^{-3} mol dm^{-3}

$10^{-14} = [H^+][10^{-3}]$

Make [H$^+$] the subject of the equation:

$$[H^+] = \frac{10^{-14}}{10^{-3}} = 10^{-11} \text{ mol dm}^{-3}$$

Step 2: Calculate the pH of the solution.

$pH = -\log[H^+]$

$-\log(10^{-11}) = 11.0$

Method 2

Just as the H$^+$ ion concentration can be expressed as a logarithmic function, so too can the ionic product of water and the hydroxide ion concentration.

$pK_w = -\log(10^{-14}) = 14.0$

$pK_w = pH + pOH$

Step 1: Calculate the pOH of 0.001 mol dm^{-3} NaOH.

$pOH = -\log(10^{-3}) = 3.0$

Step 2: Subtract the pOH from pK_w to find the pH.

$pH = pK_w - pOH$

$pH = 14.0 - 3.0 = 11.0$

Both methods are equally valid. When choosing a method, consider the method you have been taught and which method you feel most comfortable with.

> **REMEMBER:**
> Be consistent when giving your answers to a certain number of decimal places (d.p.) or significant figures (s.f.).
>
> Examiners often ask for pH to 1 d.p. or 3 s.f. You should learn the difference:
>
> pH 11.1 is both to 1 d.p. and 3 s.f.
>
> pH 0.0000123 is to 7 d.p. but has 3 s.f.

Calculating the pH of bases when you are given the mass of solute and volume of solvent

Calculate the pH when 0.28 g of KOH is dissolved in 100 cm^3.

Step 1: Find the number of moles of solute.

A_r K = 39, O = 16, H = 1

Molar mass, M, of KOH = 56

No. of moles, $n = \dfrac{m}{M} = \dfrac{0.28}{56} = 0.005$

Step 2: Find the concentration of KOH.

Concentration of OH$^- = \left(\dfrac{0.005}{100}\right) \times 1000 = 0.05 \text{ mol dm}^{-3}$

pOH $= -\log(5 \times 10^{-2}) = 1.3$ (1 d.p.)

Step 3: Find the pH.

pH $= pK_w - pOH = 14.0 - 1.3 = 12.7$ (1 d.p.)

PRACTICE QUESTIONS

1 Calculate the pH of the following solutions of KOH(aq) to 3 s.f.:

   a 0.01 mol dm^{-3}

   b 0.004 mol dm^{-3}

   c 0.000 05 mol dm^{-3}

   d 3.0 mol dm^{-3}

2 Calculate the pH of the following bases to 1 d.p.:

   a 15 g of NaOH dissolved in 75 cm^3 of water

   b 1.5 g of LiOH dissolved in 75 cm^3 of water

STRETCH YOURSELF!

How to deal with weak bases

Just as weak acids partially dissociate in solution, so do weak bases.

Pyridine, C_5H_5N, is a weak base. Its base dissociation constant, K_b, is 1.8×10^{-9}.

$$C_5H_5N + H_2O \rightleftharpoons C_5H_5NH^+ + OH^-$$

a Deduce the expressions for K_b and pK_b.

The concentration of pyridine is 1 mol dm^{-3}.

b Calculate the concentration of [OH$^-$] ions.

c Using the ionic product of water, calculate the concentration of [H$^+$] ions.

d Calculate the pH of this weak base.

e If dissociation is an endothermic process, explain in terms of Le Chatelier's principle what effect increasing temperature has on pH.

7.9 pH of buffers

Calculations involving buffer solutions

A buffer solution is a mixture of a weak acid and its conjugate base, used to keep the pH more or less constant. It cannot prevent pH changes but can minimise and prevent large changes in pH. This has applications in medicine, foodstuffs, and industry.

A typical buffer contains a weak acid and one of its salts. In order to choose the buffer with the correct pH a chemist needs to be able to calculate the acid dissociation constant, work with logarithms to calculate pK_a or pH, and find concentrations of both the weak acid and its salts.

Typical buffer solution:

$$HCOOH(aq) \rightleftharpoons HCOO^-(aq) + H^+(aq) \qquad (1)$$

$$HCOONa(aq) \rightarrow HCOO^-(aq) + Na^+(aq) \qquad (2)$$

- The weak acid is partially dissociated and the equilibrium position will change in order to minimise changes in pH and maintain the equilibrium.
- The salt is a source of base ions to ensure there is the capacity to combine with H^+ ions.

There are two methods for calculating the pH of a buffer solution.

WORKED EXAMPLE

Calculating pH of a buffer solution

Method 1: By finding the H^+ ion concentration and then the pH

Calculate the pH of a buffer solution containing $0.10 \, mol \, dm^{-3}$ methanoic acid and $0.050 \, mol \, dm^{-3}$ sodium methanoate.
K_a of methanoic acid $= 1.6 \times 10^{-4} \, mol \, dm^{-3}$

Step 1: Use the rearranged acid dissociation equation to find $[H^+]$.

$$[H^+(aq)] = K_a \times \frac{[HA]}{[A^-]}$$

$$[H^+(aq)] = 1.6 \times 10^{-4} \times \frac{(0.10)}{(0.05)} = 0.000\,320 \, mol \, dm^{-3}$$

Step 2: Find the pH.

$$pH = -\log[H^+] = -\log(0.000\,320) = 3.5 \text{ (1 d.p.)}$$

Method 2: By finding the pK_a value and logs of acid and salt concentration

This method is sometimes known as the Henderson–Hasselbalch relationship.

$$pH = pK_a + \log\left(\frac{[A^-]}{[HA]}\right)$$

pK_a of methanoic acid $= -\log(1.6 \times 10^{-4}) = 3.796$

$$pH = 3.796 + \log\left(\frac{0.05}{0.10}\right) = 3.5 \text{ (1 d.p.)}$$

> **REMEMBER:**
> $\log\left(\frac{0.05}{0.1}\right)$ is the same as $\log(0.05) - \log(0.1)$. Performing the calculation as a subtraction is much easier on a calculator.

When given the mass of solutes and the volume of the buffer solution you should do the following:

Step 1: Calculate the concentration of the acid.

Step 2: Calculate the concentration of the salt.

Step 3: Find the pH of the buffer solution.

You will need this method for Question 2 below.

PRACTICE QUESTIONS

1 Calculate the pH of the following buffers:

   a $0.20\,mol\,dm^{-3}$ phenol and $0.015\,mol\,dm^{-3}$ sodium phenoxide.
       K_a of $C_6H_5OH = 1.3 \times 10^{-10}\,mol\,dm^{-3}$

   b $0.15\,mol\,dm^{-3}$ phenol and $0.030\,mol\,dm^{-3}$ sodium phenoxide.

   c $0.10\,mol\,dm^{-3}$ benzoic acid and $0.040\,mol\,dm^{-3}$ sodium benzoate.
       K_a of $C_6H_5COOH = 6.5 \times 10^{-5}\,mol\,dm^{-3}$

   d $0.20\,mol\,dm^{-3}$ benzoic acid and $0.04\,mol\,dm^{-3}$ sodium benzoate.

   e $0.50\,mol\,dm^{-3}$ ethanoic acid and $0.025\,mol\,dm^{-3}$ sodium ethanoate.
       K_a of $CH_3COOH = 1.7 \times 10^{-5}\,mol\,dm^{-3}$

   f $0.25\,mol\,dm^{-3}$ ethanoic acid and $0.015\,mol\,dm^{-3}$ sodium ethanoate.

   g $0.30\,mol\,dm^{-3}$ hydrocyanic acid and $0.010\,mol\,dm^{-3}$ potassium cyanide.
       K_a of $HCN = 4.9 \times 10^{-10}\,mol\,dm^{-3}$

   h $0.20\,mol\,dm^{-3}$ hydrocyanic acid and $0.015\,mol\,dm^{-3}$ potassium cyanide.

   i $0.20\,mol\,dm^{-3}$ nitrous acid and $0.015\,mol\,dm^{-3}$ potassium nitrite.
       K_a of $HNO_2 = 4.5 \times 10^{-4}\,mol\,dm^{-3}$

   j $0.10\,mol\,dm^{-3}$ nitrous acid and $0.030\,mol\,dm^{-3}$ potassium nitrite.

2 Calculate the pH of these buffer solutions:

   a   A buffer solution consists of 7.00 g of ethanoic acid and 14.00 g of sodium ethanoate in 200 cm^3 of water. K_a of $CH_3COOH = 1.7 \times 10^{-5}\,mol\,dm^{-3}$

   b   A buffer solution consists of 10 g of hydrocyanic acid and 20 g of sodium cyanide in 200 cm^3 water. K_a of $HCN = 4.9 \times 10^{-10}\,mol\,dm^{-3}$

STRETCH YOURSELF!

Ratios in buffer solutions

Calculate what volumes of propanoic acid and sodium propanoate solution, both of $0.01\,mol\,dm^{-3}$ concentration, would be needed to make 200 cm^3 of pH 5.5 buffer solution. The K_a of CH_3CH_2COOH is $1.3 \times 10^{-5}\,mol\,dm^{-3}$.

Step 1:   Find [H$^+$] from the pH.

         $[H^+] = 10^{-5.5} = 3.162 \times 10^{-6}\,mol\,dm^{-3}$

Step 2:   Find the ratio of [H$^+$] : [HA] by rearranging the K_a equation.

$$\frac{[HA]}{[A^-]} = \frac{3.162 \times 10^{-6}}{1.3 \times 10^{-5}} = 0.243$$

         [HA] : 0.243[A$^-$]   Divide both sides by 0.243

         This will give you the ratio of [HA] : [A$^-$] which is 4.11 : 1.

Step 3:   Find the volume of each solution required. First, add the numbers from each side of the ratio:

         4.11 + 1 = 5.11

         Divide the volume of buffer solution required by this number.

$$\frac{200}{5.11} = 39.132$$

         Volume of [A$^-$] = 39.132 $\times$ 1 = 39.13 cm^3

         Volume of [HA] = 39.132 $\times$ 5.11 = 160.87 cm^3

8.1 Born–Haber cycles 1

Using enthalpy change data to construct Born–Haber cycles and deduce lattice enthalpy change

A Born–Haber cycle is used to calculate an unknown enthalpy change, which cannot be measured directly, from a series of known enthalpy changes.

Lattice enthalpy, ΔH_L, is not directly measurable by experiment. This can be attributed to the fact that lattice enthalpy is the formation of 1 mole of solid compound from its gaseous ions. The formation of gaseous ions is not practically achievable, but you can calculate it to give you an idea of the strength of the ionic lattice.

Chemists must be able to analyse the Born–Haber cycle diagram and deduce answers to this algebraic problem.

Before constructing a Born–Haber cycle, it is important to know the key definitions and to be able to write their equations. To complete the cycle you have to convert the elements into gaseous ions through a series of steps. It is important to know that the steps are all in standard conditions and the state symbols are important.

Atomisation enthalpy, $\Delta H_{at}^{\ominus}/kJ\,mol^{-1}$	Ionisation energy, $\Delta H_I^{\ominus}/kJ\,mol^{-1}$	Electron affinity, $\Delta H_{ea}/kJ\,mol^{-1}$	Enthalpy of formation, $\Delta H_f^{\ominus}/kJ\,mol^{-1}$	
$Li(s) \rightarrow Li(g) \quad = +159$	$Li(g) \rightarrow Li^+(g) + e^- \quad = +520$		LiCl	-409
$Na(s) \rightarrow Na(g) \quad = +109$	$Na(g) \rightarrow Na^+(g) + e^- \quad = +494$		NaCl	-411
$Mg(s) \rightarrow Mg(g) \quad = +150$	$Mg(g) \rightarrow Mg^+(g) + e^- \quad = +736$		$MgCl_2$	-642
	$Mg^+(g) \rightarrow Mg^{2+}(g) + e^- = +1450$		LiF	-612
$\frac{1}{2}F_2(g) \rightarrow F(g) \quad = +79$		$F(g) + e^- \rightarrow F^-(g) \quad = -334$	NaF	-569
$\frac{1}{2}Cl_2(g) \rightarrow Cl(g) = 121$		$Cl(g) + e^- \rightarrow Cl^-(g) = -364$	MgF_2	-1124

$\Delta H_{at}^{\ominus}$: The standard enthalpy change of atomisation is the energy required to form one mole of gaseous atoms from its element in its standard state. Chlorine, for example, as seen in the table above in standard conditions, is found as a gaseous molecule, whereas sodium is a solid.

$\Delta H_I^{\ominus}$: The first ionisation energy is the energy required to form one mole of gaseous ions of charge $+1$ from one mole of gaseous atoms. If there is a second ionisation energy it is the loss of a further electron to form a $+2$ ion.

ΔH_{ea}: The electron affinity is the energy lost in forming one mole of gaseous ions of charge -1 from one mole of gaseous atoms.

$\Delta H_f^{\ominus}$: The standard enthalpy change of formation is for formation of one mole of a compound from an element in its standard state.

For example: $Mg(s) + F_2(g) \rightarrow MgF_2(s)$

The standard lattice enthalpy can be calculated by subtracting the standard enthalpy of formation from all the previous steps.

> **REMEMBER:** When drawing a Born–Haber cycle you always start with the elements in their standard states. So it is important you leave room for the ΔH_L and ΔH_f.

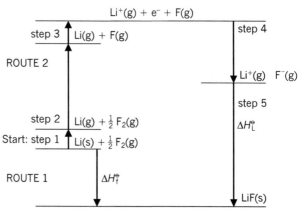

WORKED EXAMPLE

Drawing Born–Haber cycles for Group 2 elements

Born–Haber cycles of Group 2 elements have three differences from the Group 1 elements. For example, $CaCl_2$:

1   All Group 2 elements form ions of charge +2 so you have to include both ionisation energies.

2   There are 2 moles of Cl so the atomisation of chlorine has to be doubled.

3   There are 2 moles of Cl so the electron affinity has to be doubled too.

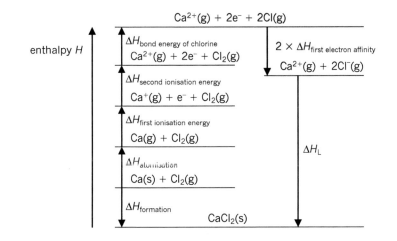

PRACTICE QUESTION

1   Construct Born–Haber cycles for each of these compounds:

   a   LiCl

   b   NaF

   c   NaCl

   d MgF_2

STRETCH YOURSELF!

Construct a Born–Haber cycle for Al_2O_3

Draw a Born–Haber cycle for Al_2O_3.

Note:

* There are 2 moles of aluminium so the atomisation has to be doubled.

* There are 3 moles of oxygen so the atomisation has to be tripled.

* The three ionisation energies for aluminium are all doubled.

* The two electron affinities for oxygen are all tripled.

8.2 Born–Haber cycles 2

Carrying out calculations using Born–Haber cycles to find an unknown lattice enthalpy

The lattice energy of an ionic crystal is the heat of formation for 1 mole of an ionic compound from gaseous ions under standard conditions.

Example: $Li^+(g) + F^-(g) \rightarrow Li^+F^-(s)$

The lattice enthalpy is helpful in discussing the structure, bonding, and reactions of this compound.

Unfortunately, lattice enthalpy cannot be determined directly, but values can be obtained by using a Born–Haber energy cycle, such as the one for LiF in the previous topic.

WORKED EXAMPLE

Using a Born–Haber cycle to calculate the lattice enthalpy of lithium fluoride

Hess's law states that the total enthalpy change is independent of the reaction route taken.

This principle can be used to find an unknown enthalpy change in a Born–Haber cycle. If you look at the diagram below you will see there are two routes to produce an ionic compound from elements in their standard states. Applying Hess's Law means that both routes have the same enthalpy change.

Route 1 = Route 2

From the diagram this can be expressed as:

Sum of anticlockwise enthalpies = sum of clockwise enthalpies.

$$\Delta H_1 = \Delta H_2 + \Delta H_3 + \Delta H_4 + \Delta H_5 + \Delta H_6$$

ΔH_1 is $\Delta H_f(LiF)$

ΔH_6 is $-\Delta H_L(LiF)$

$$\Delta H_6 = \Delta H_1 - \Delta H_2 - \Delta H_3 - \Delta H_4 - \Delta H_5$$

$$\Delta H_f(LiF) = \Delta H_a(Li) + \Delta H_a(F) + \Delta H_i(Li) + \Delta H_{ea}(F) - \Delta H_L(LiF)$$

$$\Delta H_L(LiF) = -\Delta H_f(LiF) - (\Delta H_a(Li) + \Delta H_a(F) + \Delta H_i(Li) + \Delta H_{ea}(F))$$

$$\Delta H_L = -612 - (159 + 79 + 520 - 334) = -1036\,kJ\,mol^{-1}$$

> **NOTE:** ΔH_L is always exothermic and therefore negative. Remember to check your calculations.

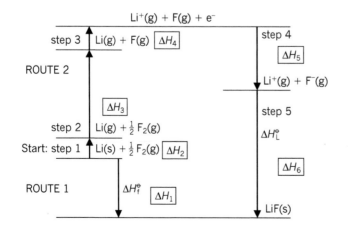

> **REMEMBER:** If you go the opposite route to the direction of an arrow it is a minus, and a minus of a minus is a plus!

PRACTICE QUESTIONS

Reaction route taken	$\Delta H / \text{kJ mol}^{-1}$
a $K(s) + \frac{1}{2}Br_2(l) \rightarrow K^+Br^-(s)$	-392
b $K(s) \rightarrow K(g)$	$+90$
c $K(g) \rightarrow K^+(g) + e^-$	$+420$
d $\frac{1}{2}Br_2(l) \rightarrow Br(g)$	$+112$
e $Br(g) + e^- \rightarrow Br^-(g)$	-342

1 For A to E, identify the name of each enthalpy change.

2 Give an equation for the lattice enthalpy, ΔH_L, of KBr.

3 Construct a Born–Haber cycle for the formation of KBr.

4 Determine the lattice enthalpy of KBr.

5 State how you expect the lattice enthalpy of KBr to compare to NaBr and KF. Give reasons for your answers.

STRETCH YOURSELF!

Calculate the enthalpy of formation of a hypothetical compound

a Explain what is meant by the terms (i) ionisation energy, (ii) atomisation energy, (iii) lattice enthalpy, and (iv) enthalpy of formation.

Magnesium chloride has the formula $MgCl_2$, that is, $Mg^{2+}(Cl^-)_2$, and a standard enthalpy of formation of $-642\ \text{kJ mol}^{-1}$.

b Draw the Born–Haber cycle for $MgCl_2$.

Atomisation enthalpy of magnesium $= +150\ \text{kJ mol}^{-1}$

Atomisation enthalpy of chlorine $= +121\ \text{kJ mol}^{-1}$

First ionisation energy of magnesium $= +736\ \text{kJ mol}^{-1}$

First electron affinity of chlorine $= -364\ \text{kJ mol}^{-1}$

c The lattice enthalpy of $MgCl_2$ is $-2492\ \text{kJ mol}^{-1}$ and the enthalpy of formation of $MgCl_2$ is $-642\ \text{kJ mol}^{-1}$. Calculate the second ionisation energy of magnesium.

8.3 Enthalpy of solution

Using enthalpy changes of solution to calculate Born–Haber cycles

The standard enthalpy change of solution, $\Delta H^{\ominus}_{sol}$, is the enthalpy change that takes place when one mole of a compound completely dissolves in water under standard conditions.

The enthalpy change of solution can be determined by measuring the temperature change.

When an ionic solid dissolves in water, two things happen:

1 The ionic lattice breaks down into gaseous ions, an endothermic process. (This is the exact opposite of the process involved in lattice enthalpy, $-\Delta H_L$.)

2 These ions are hydrated when they form bonds with water, ΔH_{hyd}, an exothermic process.

$$\Delta H^{\ominus}_{sol} = \Sigma \Delta H_{hyd} - \Delta H_L$$

The enthalpy of solution or the lattice enthalpy can be calculated if you know the other two by using a Born–Haber cycle and Hess's law.

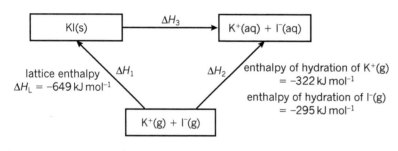

✓ WORKED EXAMPLE

How to calculate the enthalpy change of solution of KI

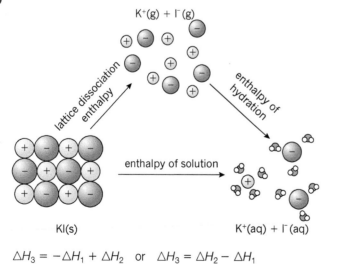

$\Delta H_3 = -\Delta H_1 + \Delta H_2$ or $\Delta H_3 = \Delta H_2 - \Delta H_1$

$\Delta H_3 = (-322 + -295) - (-649) = -617 + 649 = +32 \text{ kJ mol}^{-1}$

Dissolving potassium iodide must be an endothermic process and result in a temperature drop in solution.

> **REMEMBER:** Enthalpies with highly endothermic values for ΔH_{sol} are insoluble. If it is slightly endothermic it can be soluble if compensated for by an increase in entropy.

Calculating lattice enthalpy from enthalpy change of solution and enthalpy of hydration

In order to find the ΔH_L of KCl, you need to use Hess's law, which states that enthalpy change is independent of the reaction route taken.

So you just need to take an alternative route to find the enthalpy of turning from gaseous ions to an ionic lattice.

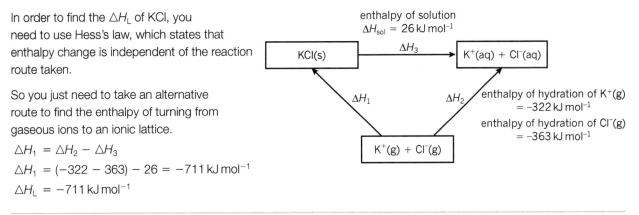

$\Delta H_1 = \Delta H_2 - \Delta H_3$

$\Delta H_1 = (-322 - 363) - 26 = -711 \text{ kJ mol}^{-1}$

$\Delta H_L = -711 \text{ kJ mol}^{-1}$

PRACTICE QUESTIONS

1   Draw and label an enthalpy cycle diagram for the enthalpy change of solution of LiCl(s).

2   Calculate the enthalpy change of solution for LiCl from the data:
$\Delta H_L(\text{LiCl}) = +853 \text{ kJ mol}^{-1}$, $\Delta H_{hyd}(\text{Li}^+) = -520 \text{ kJ mol}^{-1}$,
$\Delta H_{hyd}(\text{Cl}^-) = -363 \text{ kJ mol}^{-1}$

3   Draw and label an enthalpy cycle diagram for the enthalpy change of solution of NaCl(s).

4   Calculate the enthalpy change of solution for NaCl from the data:
$\Delta H_L(\text{NaCl}) = +786 \text{ kJ mol}^{-1}$, $\Delta H_{hyd}(\text{Na}^+) = -406 \text{ kJ mol}^{-1}$,
$\Delta H_{hyd}(\text{Cl}^-) = -363 \text{ kJ mol}^{-1}$

5   Draw and label an enthalpy cycle diagram for the enthalpy change of solution of $MgCl_2$.

6   Calculate the enthalpy change of solution for $MgCl_2$ from the data:
$\Delta H_L(\text{MgCl}_2) = +2526 \text{ kJ mol}^{-1}$, $\Delta H_{hyd}(\text{Mg}^{2+}) = -1920 \text{ kJ mol}^{-1}$,
$\Delta H_{hyd}(\text{Cl}^-) = -363 \text{ kJ mol}^{-1}$

7   Draw and label an enthalpy cycle diagram to find ΔH_L of AgF.

8   Calculate the lattice enthalpy of of AgF given the data:
$\Delta H_{sol}(\text{AgF}) = -10 \text{ kJ mol}^{-1}$, $\Delta H_{hyd}(\text{Ag}^+) = -464 \text{ kJ mol}^{-1}$,
$\Delta H_{hyd}(\text{F}^-) = -506 \text{ kJ mol}^{-1}$

STRETCH YOURSELF!

Why is it less soluble?

Salt	Lattice enthalpy / kJ mol^{-1}	Ions	Hydration enthalpy / kJ mol^{-1}
BaF$_2$(s)	−2352	Ba^{2+}	−1275
		F$^-$	−506
BaCl$_2$(s)	−2056	Ba^{2+}	−1275
		Cl$^-$	−363

Construct enthalpy cycle diagrams and find the enthalpy of solution of BaCl$_2$ and BaF$_2$. Explain why BaF$_2$ is less soluble than BaCl$_2$.

8.4 Factors that affect lattice enthalpies

Explaining qualitatively how ionic charge and ionic radius affect lattice enthalpy and enthalpy of hydration

There are two factors that determine a salt's lattice enthalpy:

- the size of the ion (ionic radius)
- the charge on the ions.

A smaller ionic radius would mean a higher charge density. This would mean a greater attraction for water molecules and a more exothermic enthalpy of hydration.

A bigger charge on the ion would mean greater attraction for the water molecules and therefore a more exothermic enthalpy of hydration.

$$\Delta H_{sol}^{\ominus} = \Sigma \Delta H_{hyd} - \Delta H_L$$

A smaller ionic radius and bigger charge also means a larger lattice enthalpy because of greater attraction between the ions.

As both $\Sigma \Delta H_{hyd}$ and ΔH_L increase it makes it difficult to predict the effect of these factors on $\Delta H_{sol}^{\ominus}$

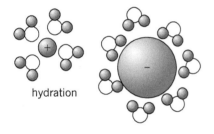

hydration

The diagram shows the electrostatic attraction between ions and polar water molecules.

WORKED EXAMPLE

Comparing two lattice enthalpies

Why is the lattice enthalpy of sodium iodide smaller than the lattice enthalpy of sodium fluoride?

$\Delta H_L(\text{NaI}) = -704\,\text{kJ mol}^{-1}$

$\Delta H_L(\text{NaF}) = -923\,\text{kJ mol}^{-1}$

A fluoride ion has a smaller ionic radius than an iodide ion. Therefore, sodium fluoride has a higher lattice enthalpy and a more exothermic enthalpy of hydration than sodium iodide.

> **REMEMBER:** The definition of lattice enthalpy change is the enthalpy change when 1 mole of an ionic compound is formed from its ions under standard conditions.
>
> $A^+(g) + B^-(g) \rightarrow AB(s)$
>
> Lattice enthalpy values are exothermic because new bonds are made.

WORKED EXAMPLE

Predicting lattice enthalpies

The four lattice enthalpy values are $-821\,kJ\,mol^{-1}$, $-682\,kJ\,mol^{-1}$, $-2440\,kJ\,mol^{-1}$, and $-2957\,kJ\,mol^{-1}$.

The four ionic compounds with these values are $MgBr_2$, KBr, MgF_2, and KF.

Match the values and justify your reasons.

Ionic compound	Lattice enthalpy / kJ mol^{-1}
MgF_2	−2957
$MgBr_2$	−2440
KF	−821
KBr	−682

Mg^{2+} has a smaller ionic radius and greater charge than K^+ so must have a more exothermic hydration energy and a larger lattice enthalpy.

F^- has a smaller ionic radius than Br^- so must have a more exothermic hydration energy and a larger lattice enthalpy.

PRACTICE QUESTIONS

1   For each of the following pairs of ionic compounds, state with reasons which you would expect to have the higher lattice enthalpy.

   a   NaCl and KBr

   b   CaO and MgO

   c K_2S and Na_2S

   d $MgCl_2$ and RbCl

   e $MgSO_4$ and MgO

2   For the thermal decomposition of Group 2 carbonates:

   a   Give a balanced equation for the decomposition of a Group 2 carbonate, XCO_3.

   b   Describe the trend in stability for Group 2 carbonates from $MgCO_3$ to $BaCO_3$.

   c   Explain the trend you have described.

STRETCH YOURSELF!

Solubilities of hydroxides and sulfates

The solubilities of Group 2 hydroxides increase down the group but the solubilities of sulfates decrease. Suggest a reason for these opposing trends in terms of lattice enthalpies and enthalpies of hydration.

9 ELECTROPOTENTIALS AND PREDICTING REACTIONS

9.1 Calculating the cell potential

Calculating a standard cell potential by combining two standard electrode potentials

In an electrochemical cell there is a redox reaction where a transfer of electrons occurs between one metal and another.

The electron transfer process can be predicted by knowing the standard electrode potentials. Using these you can find the species that would be oxidised and the species that would be reduced and determine the potential difference between them.

$$Zn(s) \rightarrow Zn^{2+}(aq) + 2e^-$$

$$Cu^{2+}(aq) + 2e^- \rightarrow Cu(s)$$

Electromotive force (emf) is a measurement of the voltage across the cell and emf under standard temperature and conditions of 298 K and 100 kPa is given the symbol $E^\ominus$.

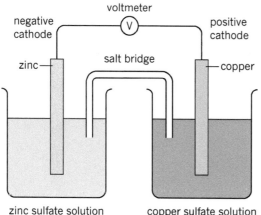

zinc sulfate solution   copper sulfate solution

Standard electrode reduction potentials

Half cell	Half reaction	$E^\ominus$/V
$Na^+(aq)\,\vert\,Na(s)$	$Na^+(aq) + e^- \rightarrow Na(s)$	−2.71
$Al^{3+}(aq)\,\vert\,Al(s)$	$Al^{3+}(aq) + 3e^- \rightarrow Al(s)$	−1.67
$Fe^{2+}(aq)\,\vert\,Fe(s)$	$Fe^{2+}(aq) + 2e^- \rightarrow Fe(s)$	−0.44
$2H^+(aq), H_2(g)\,\vert\,Pt$	$2H^+(aq) + 2e^- \rightarrow H_2(g)$	0.00
$Br_2(aq), Br^-(aq)\,\vert\,Pt$	$Br_2(aq) + 2e^- \rightarrow 2Br^-(aq)$	+1.09
$Cl_2(aq), Cl^-(aq)\,\vert\,Pt$	$Cl_2(aq) + 2e^- \rightarrow 2Cl^-(aq)$	+1.51

NOTE: A standard electrode potential cannot be determined by itself. It has to be compared with a known reference. Platinum (Pt) is used as an electrode in cases where it is not possible to build an electrode out of the species. This is because platinum is inert and does not interfere with the electrochemical reaction.

REMEMBER: The more positive the number and the lower down the table the more likely a species is to be reduced.

WORKED EXAMPLE
Writing conventional cell diagrams for electrochemical cells

Rather than drawing a completed diagram of an electrochemical cell each time, there is a convenient shorthand diagram.

In the cell shown on the previous page the zinc is oxidised and the copper is reduced.

Shorthand diagram:

$Zn(s)|Zn^{2+}(aq)||Cu^{2+}(aq)|Cu(s)$

Standard electrode potentials:

$Zn^{2+}|Zn = -0.76\,V$

$Cu^{2+}|Cu = +0.34\,V$

$E^{\ominus}_{cell} = E^{\ominus}_{\text{right-hand electrode}} - E^{\ominus}_{\text{left-hand electrode}} = 0.34 - -0.76 = +1.10\,V$

> **REMEMBER:** When you draw electrochemical cells or write the shorthand diagram, the more positive electrode must always be on the right.

PRACTICE QUESTION

1 For each combination of the half cells listed below:

   a Give the shorthand for the electrochemical cell.

   b Identify the electrode being reduced and the electrode being oxidised.

   c Calculate the $E^{\ominus}_{cell}$.

     i Na and Cl_2

     ii Na and Br_2

     iii Fe and H_2

     iv Na and Fe

     v Al and Fe

     vi Br_2 and Cl_2

STRETCH YOURSELF!
Hydrogen electrode

If an electrochemical cell is left on for a long time, does the pH change? Why?

Take the measurements of the standard electrode potentials of Br_2 and Fe as two examples.

$2H^+(aq), H_2(g)|Pt$	$2H^+(aq) + 2e^- \rightarrow H_2(g)$	$E^{\ominus} = 0.00\,V$
$Br_2(aq), Br^-(aq)|Pt$	$Br_2(aq) + 2e^- \rightarrow 2Br^-(aq)$	$E^{\ominus} = +1.09\,V$

Using the redox potentials determine which species, Br_2 or H^+, is being oxidised or reduced.

- Combine the two half equations for a full ionic equation.
- Determine whether the concentration of hydrogen ions will increase or decrease.

     $Fe^{2+}(aq)|Fe(s)\ Fe^{2+}(aq) + 2e^- \rightarrow Fe(s)\quad E^{\ominus} = -0.44$

- Use the redox potentials to determine which species, iron or hydrogen, is oxidised or reduced.
- Combine the two half equations for a full ionic equation.
- If the hydrogen ion increases, explain how this will affect the pH.
- Suggest what will happen to pH in both electrochemical cells.

9.2 Redox equations

Carrying out structured calculations on species that are either oxidised or reduced

Redox titrations can often be difficult to balance because they can contain a number of complex or spectator ions. By converting the equation into two ionic half equations, the equation can be balanced through simple use of simultaneous equations.

You are already familiar with acid and base titrations, which involve the transfer of protons from acids. Redox titrations are similar but involve the transfer of electrons.

$$Cr_2O_7{}^{2-}(aq) + 14H^+(aq) + 6e^- \rightarrow 2Cr^{3+}(aq) + 7H_2O(l)$$

Dichromate is an excellent oxidising agent. An additional indicator is not needed as the end point can be determined by the colour change from orange to green.

Dichromate can be titrated against a reducing agent and, in the same way as in an acid–base titration, can be measured quantitatively.

Follow these steps to balance a redox equation using the ionic half equation method:

1 Convert the unbalanced equation into an ionic equation.

2 Assign oxidation numbers to each species.

3 Write ionic half equations — one for the species that has been oxidised and one for the species that has been reduced. Then balance all atoms, with the exception of the oxygen and hydrogen atoms.

4 Balance all the oxygen then hydrogen atoms.

5 Balance the ionic charge by adding electrons to both half equations.

6 Treat as simultaneous equations. Balance electron loss with electron gain for both half equations and then add both half equations.

7 Change back to molecular form by adding spectator ions.

 WORKED EXAMPLE

$HBr + H_2SO_4 \rightarrow Br_2 + SO_2 + H_2O$

Step 1: $H^+ Br^- + H^+ HSO_4{}^- \rightarrow Br_2 + SO_2 + H_2O$

Step 2: $Br^- = -1 \quad S = +6 \rightarrow Br = 0 \quad S = +4$

Bromine has been oxidised.

Sulfur has been reduced.

Step 3: $2Br^- \rightarrow Br_2$

$HSO_4{}^- + H^+ \rightarrow SO_2 + H_2O$

Bromine atoms balance.

Sulfur atoms balance.

Step 4 and **Step 5:**

$2Br^- \rightarrow Br_2 + 2e^-$

$HSO_4{}^- + 2e^- + 3H^+ \rightarrow SO_2 + 2H_2O$

Oxygen atoms, hydrogen atoms, and charge all balance.

> **REMEMBER:** Learn the mnemonic OIL RIG:
>
> Oxidation is loss (*of electrons*).
>
> Reduction is gain (*of electrons*).
>
> Oxidation can also be defined as an increase in oxidation number.
>
> Conversely, reduction is a decrease in oxidation number.

Step 6: Add the two half equations and cancel the electrons.

$2Br^- + HSO_4^- + \cancel{2e^-} + 3H^+ \rightarrow Br_2 + \cancel{2e^-} + SO_2 + 2H_2O$

$2Br^- + 3H^+ + HSO_4^- \rightarrow Br_2 + SO_2 + 2H_2O$

Step 7: $2HBr + H_2SO_4 \rightarrow Br_2 + SO_2 + 2H_2O$

Equation is now in the molecular form.

PRACTICE QUESTIONS

1 Construct redox equations for the following pairs of half equations:

   a $Mg \rightarrow Mg^{2+} + 2e^-$, $Cu^{2+} + 2e^- \rightarrow Cu$

   b $Al \rightarrow Al^{3+} + 3e^-$, $F_2 + 2e^- \rightarrow 2F^-$

   c $Na \rightarrow Na^+ + e^-$, $2H^+ + 2e^- \rightarrow H_2$

   d $H_2 \rightarrow 2H^+ + 2e^-$, $O_2 + 4e^- \rightarrow 2O^{2-}$

2 Give the oxidation number of the following elements in complex ions:

   a Mn in MnO_4^-

   b Cr in $Cr_2O_7^{2-}$

   c Cu in Cu_2O

   d Cr in $Cr_2(SO_4)_3$

3 Give the balanced redox equations for the following:

   a $Cu + H^+ + NO_3^- \rightarrow Cu^{2+} + NO + H_2O$

   b $Na + Cr_2O_7^{2-} + H^+ \rightarrow Na^+ + Cr^{3+} + H_2O$

STRETCH YOURSELF!

Redox titrations

Acidified MnO_4^- oxidises Fe^{2+} to Fe^{3+}, and is in turn reduced from MnO_4^- to Mn^{2+}.

In a titration $24.8\,cm^3$ of acidified $0.02\,mol\,dm^{-3}$ MnO_4^- is titrated against $20\,cm^3$ of Fe^{2+} solution of unknown concentration.

1 Give the balanced half equations.

2 Give the balanced redox equation.

3 Calculate the concentration of Fe^{2+} solution.

4 Given the half-reactions:

   $H_2C_2O_4(aq) \rightleftharpoons 2CO_2(g) + 2H^+(aq) + 2e^-$

   and

   $MnO_4^- + 8H^+ + 5e^- \rightarrow Mn^{2+} + 4H_2O$

   a Give the balanced redox equation for MnO_4^- ions oxidising ethanedioic acid.

   b $1.50\,g$ of ethanedioic acid crystals, $H_2C_2O_4.2H_2O$, was made up to $250\,cm^3$ of aqueous solution and $25.0\,cm^3$ of this solution needed $12.50\,cm^3$ of a potassium manganate(VII) solution for oxidation. Calculate the concentration in $mol\,dm^{-3}$ of MnO_4^-.

9.3 Redox titrations

Carrying out structured calculations involving thiosulfate and iodine molecules

Sodium hypochlorite, $NaClO$, is the active ingredient in bleach. Its concentration can be determined by a multi-stage redox titration.

1   An excess of KI is added to $5\,cm^3$ of bleach. The iodide is oxidised to iodine.

2   A small amount of starch solution is added as an indicator. Iodine goes dark blue.

3   The iodine is then titrated against sodium thiosulfate, $Na_2S_2O_3$, where the iodine is reduced back to iodide.

4   The volume of sodium thiosulfate is recorded and the number of moles is calculated.

5   The number of moles of iodine is then calculated.

6   This indirectly tells us the number of moles of hypochlorite and enables us to calculate the hypochlorite concentration of bleach.

$$ClO^-(aq) + 2I^-(aq) + 2H^+(g) \rightarrow Cl^-(aq) + I_2(aq) + H_2O(l)$$
$$I_2(aq) + starch \rightarrow 2I^-[starch]^{2+}\ complex$$
$$2S_2O_3^{2-}(aq) + I_2(aq) \rightarrow 2I^-(aq) + S_4O_6^{2-}(aq)$$

WORKED EXAMPLE

How to calculate the hypochlorite concentration of bleach

$5\,cm^3$ of bleach A was diluted to $25.0\,cm^3$ with deionised water. An excess of iodide ions was added, reducing the hypochlorite ions. The iodine was then titrated against $0.01\,mol\,dm^{-3}$ $Na_2S_2O_3$ and $24.00\,cm^3$ of $Na_2S_2O_3$ solution was added.

Step 1:   Find the number of moles of thiosulfate added.

$$Concentration,\ c = \frac{number\ of\ moles,\ n}{volume,\ V\ (in\ dm^3)}$$

$$n = c \times V\ (in\ dm^3) = \frac{c \times V\ (in\ cm^3)}{1000}$$

$$n = \frac{0.01 \times 24.00}{1000}$$

$$S_2O_3^{2-} = 0.00024\,mol$$

Step 2:   Calculate the number of moles of iodide.

From the third equation above: $S_2O_3^{2-}:I^-$ 2:2 mol ratio 1:1 mol ratio

Iodide $I^- = 0.00024\,mol$

Step 3:   Determine the number of moles of hypochlorite.

From the first equation above: $I^-:ClO^-$ 2:1 mol ratio

$ClO^- = 0.00012\,mol$

Step 4: Find the concentration of hypochlorite ions.

$$c = \frac{n}{V \text{ (in dm}^3)} = \frac{n}{V \text{ (in cm}^3)} \times 1000$$

$$c = \frac{0.000\,12}{5} \times 1000 = 0.024 \, \text{mol dm}^{-3}$$

Concentration of bleach A = $0.024 \, \text{mol dm}^{-3}$

> **Note:** to convert volume in cm^3 into dm^3 use:
>
> Volume (in dm^3) = $\dfrac{\text{volume (in cm}^3)}{1000}$

> **REMEMBER:**
> Concentration is measured in mol dm^{-3}, so you must convert volumes from cm^3 to dm^3 as part of your calculation.
>
> It is good practice to do this in the calculation by dividing by 1000.

PRACTICE QUESTIONS

1   Two types of bleaches are titrated. Bleach B requires $17.0 \, \text{cm}^3$ of thiosulfate. Bleach C requires $29.5 \, \text{cm}^3$.

   Find the concentrations of bleaches B and C.

2   The active ingredient of bleach is sodium hypochlorite. In trials, bleach A is found to be superior but it doesn't have the highest hypochlorite concentration. Name two other factors that need to be considered.

STRETCH YOURSELF!

Find the concentration of thiosulfate

1   The analysis on bleach A is repeated by another team. They used $5 \, \text{cm}^3$ of bleach A, but required only $3.50 \, \text{cm}^3$ of thiosulfate.

   a Determine the concentration of the thiosulfate solution they used. The concentration of bleach was $0.024 \, \text{mol dm}^{-3}$.

   b Out of the two concentrations of thiosulfate used, identify which was most suitable. Justify your answer.

2   The commercial reaction used to prepare bleach involves chlorine dissolving in NaOH.

   $$Cl_2 + 2OH^- \rightleftharpoons Cl^- + ClO^- + H_2O$$

   a Identify the change in oxidation numbers of Cl in the reaction.

   b Define disproportionation.

   c Suggest why this reaction is important for the activity of bleach.

1 The contact process is used for the industrial production of sulfuric acid. The main reaction in the process is the production of sulfur trioxide, which is an exothermic process. The temperature is 750 °C and the pressure is 400 kPa.

$$2 SO2(g) + O2(g) \rightleftharpoons 2 SO3(g) : \Delta H = -197 \, kJ \, mol^{-1}$$

At equilibrium there are 40 moles of SO_2, 20 moles of O_2, and 960 moles of SO_3.

a Give the expression for K_p.

b Calculate K_p.

c Describe the effects on the equilibrium if temperature is increased and if pressure is decreased.

d Explain how this would affect the value of K_p.

e Determine how many moles of each reactant were present prior to equilibrium.

2 When a sample of 0.0450 moles of dinitrogen tetroxide dissociated at 25 °C in a container of volume 2 dm^3, 0.0600 moles of N_2O_4(g) remain in the equilibrium mixture.

$$N_2O_4(g) \rightleftharpoons 2NO_2(g)$$

a Give the expression for K_c.

b Determine the units for K_c.

c Calculate K_c.

d Calculate the percentage yield of the reaction.

3 Nitric acid is a strong acid. Sodium hydroxide is a strong alkali.

Hydrofluoric acid, HF, has an acid dissociation constant, K_a, of 7.4×10^{-4}.

a Calculate the pH of a nitric acid solution of concentration 0.2 mol dm^{-3}.

b Calculate the pH of a sodium hydroxide solution of 0.2 mol dm^{-3}.

c Calculate the pH of a hydrofluoric acid solution of concentration 0.2 mol dm^{-3}.

d Calculate the pH of a solution of 4 g of NaF in 100 cm^3 of hydrofluoric acid of concentration 0.2 mol dm^{-3}.

e Calculate the pH of the solution when 25 cm^3 of NaOH of 0.2 mol dm^{-3} concentration is added to 25 cm^3 of 0.2 mol dm^{-3} HF.

4 The next question is about the ionic crystal caesium oxide.

a Name the energy, ΔH, in each of the following processes.

i $2Cs^+(g) + O_2^-(g) \rightarrow Cs_2O(s)$

ii $O(g) + 1e^- \rightarrow O^-(g)$

iii $2Cs(s) + \frac{1}{2}O_2(g) \rightarrow Cs_2(s)$

b Draw a Born–Haber cycle for the formation of caesium oxide.

c Use the following data to calculate the lattice enthalpy of caesium oxide.

Enthalpy of formation of caesium oxide = $-233 \, kJ \, mol^{-1}$

Enthalpy of sublimation of Cs = $+78 \, kJ \, mol^{-1}$

First ionisation energy of Cs = $+375 \, kJ \, mol^{-1}$

Enthalpy of dissociation of O_2(g) = $+494 \, kJ \, mol^{-1}$

First electron affinity of O = $-141 \, kJ \, mol^{-1}$

Second electron affinity of O = $+ 845 \, kJ \, mol^{-1}$

5 Electropotentials can be used to determine the emf of a cell. Below is a list of electropentials.

Electropotential chart	$E^{\ominus}/V$
$Cl_2(g) + 2e^- \rightarrow 2Cl^-(aq)$	+1.36
$Br_2(l) + 2e^- \rightarrow 2Br^-(aq)$	+1.07
$NO_3^-(aq) + 3H^+(aq) + 2e^- \rightarrow HNO_2(aq) + H_2O(1)$	+0.94
$Fe^{3+}(aq) + e^- \rightarrow Fe^{2+}(aq)$	+0.77
$I_2(aq) + 2e^- \rightarrow 2I^-(aq)$	+0.54
$VO^{2+}(aq) + 2H^+(aq) + e^- \rightarrow V^{3+}(aq) + H_2O(1)$	+0.34
$V^{3+}(aq) + e^- \rightarrow V^{2+}(aq)$	−0.26
$Fe^{2+}(aq) + 2e^- \rightarrow Fe(s)$	−0.44
$Mg^{2+}(aq) + 2e^- \rightarrow Mg(s)$	−2.37

a Define oxidising agent and reducing agent.

b Identify the strongest oxidising agent in the half equations above.

c Calculate the emf of this cell:

$Mg(s)|Mg^{2+}(aq)||Fe^{2+}(aq)|Fe(s)$

d Identify which electrode in the above cell is the anode.

e Give an equation to represent the reaction in the cell.

f Give the shorthand electrochemical cell between Fe and V.

g Calculate the emf of this cell.

6 A titration was carried out with $20.0\,cm^3$ of $0.1\,mol\,dm^{-3}$ $KMnO_4$ solution and $24.0\,cm^3$ of vanadium(II) sulfate solution of unknown concentration.

$MnO_4-(aq) + 8H^+(aq) + 5e^- \rightarrow Mn^{2+}(aq) + 4H2O(l)$

$V^{3+}(aq) + e^- \rightarrow V^{2+}(aq)$

a Write the shorthand diagram for the electrochemical cell.

b Give a fully balanced redox equation.

c Calculate the concentration of V^{2+}.

SPECTROSCOPY DATA TABLES

NMR chemical shifts relative to trimethylsilane, TMS

Chemical shifts are typical values and can vary slightly depending on the solvent, concentration, and substituents.

Type of proton	Chemical shift, δ/ppm
$R—CH_3$	0.7–1.6
$N—H$ $R—OH$	1.0–5.5*
$R—CH_2—R$	1.2–1.4
R_3CH	1.6–2.0
$H_3C—C$, $RCH_2—C$, $R_2CH_2—C$ (with =O)	2.0–2.9
CH_3, CH_2R, CHR_2 (benzene rings)	2.3–2.7
$N—CH_3$ $N—CH_2R$ $N—CHR_2$	2.3–2.9
$O—CH_3$ $O—CH_2R$ $O—CHR_2$	3.3–4.3
Br or Cl—CH_3 Br or Cl—CH_2R Br or Cl—CHR_2	3.0–4.2
OH	4.5–10.0*
$—CH = CH—$	4.5–6.0
$—C$ (=O) NH_2, $—C$ (=O) $HN—$	5.0–12.0*
H	6.5–8.0
$—C$ (=O) H	9.0–10
$—C$ (=O) $O—H$	11.0–12.0*

* OH and NH chemical shifts are very variable (sometimes outside these limits) and are often broad. Signals are not usually seen as split peaks.

Type of carbon	Chemical shift, δ / ppm
C—**C** (alkanes)	10–35
	20–30
C—Cl or **C**—Br	30–70
C—N (amines)	35–60
C—OH	50–65
C＝**C** (alkenes)	115–140
aromatic	125–150
carbonyl (ester, carboxylic acid, amide)	160–185
carbonyl (aldehyde, ketone)	190–220

Infra red spectroscopy characteristic peaks

Bond	Location	Wavenumber / cm⁻¹
C—O	alcohols, esters, carboxylic acids	1000–1300
C＝O	aldehydes, ketones, carboxylic acids, esters, amides	1640–1750
C—H	organic compound with a C—H bond	2850–3100
O—H	carboxylic acids	2500–3300 (very broad)
N—H	amines, amides	3200–3500
O—H	alcohols, phenols	3200–3550 (broad)

GRAPHS

Why graphs are useful

The results of experiments are often recorded in tables but then displayed as graphs. Graphs allow you to see patterns in the data and spot any anomalous results (results that that do not fit the pattern) more easily than when the data is presented in tables.

Graph drawing rules

- The axes of graphs need to take up at least half of the available space on the graph paper (if it does not, try adjusting the scale, e.g., go up in fives rather than tens).
- The axes must be labelled and units have to be included, for example, mass / g.
- Make the values in the axes go up in ones, twos, fives, tens, or twenties. This makes plotting the data much easier.
- Data points should be drawn as crosses so that the points can be seen even after the line of best fit has been added.
- The data should be spread over the graph as much as possible. Whilst some graphs may include an origin, 0,0, on other graphs it may be more sensible to start the axes at a different value.
- A line of best fit can be added to graphs. Use a transparent ruler and draw the line so that the points are evenly distributed on either side of the line. Take care not to simply join the first and last data point. Lines of best fit can be straight or curved lines.

 ## WORKED EXAMPLE

Plot a graph of temperature against rate of reaction.

Temperature/°C	Time/s	Rate of reaction/s^{-1}
20	25	0.04
30	20	0.05
40	16	0.06
50	11	0.09
60	7	0.14

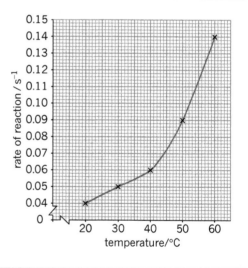

 ## PRACTICE QUESTION

1 Plot a graph of the mass of calcium carbonate against the volume of carbon dioxide made.

Mass of calcium carbonate/g	Volume of carbon dioxide made/cm^3
0.05	12
0.10	23
0.15	35
0.20	46
0.25	58
0.30	70

Periodic table

1 H 1.007 Hydrogen																	2 He 4.002 Helium
3 Li 6.941 Lithium	4 Be 9.012 Beryllium											5 B 10.811 Boron	6 C 12.011 Carbon	7 N 14.001 Nitrogen	8 O 15.999 Oxygen	9 F 18.998 Fluorine	10 Ne 20.180 Neon
11 Na 22.990 Sodium	12 Mg 24.305 Magnesium											13 Al 26.982 Aluminium	14 Si 28.085 Silicon	15 P 30.974 Phosphorus	16 S 32.065 Sulfur	17 Cl 35.453 Chlorine	18 Ar 39.948 Argon
19 K 39.098 Potassium	20 Ca 40.078 Calcium	21 Sc 44.945 Scandium	22 Ti 47.867 Titanium	23 V 50.941 Vanadium	24 Cr 51.996 Chromium	25 Mn 54.938 Manganese	26 Fe 55.845 Iron	27 Co 58.933 Cobalt	28 Ni 58.693 Nickel	29 Cu 63.546 Copper	30 Zn 65.409 Zinc	31 Ga 69.723 Gallium	32 Ge 72.64 Germanium	33 As 74.922 Arsenic	34 Se 78.96 Selenium	35 Br 79.904 Bromine	36 Kr 83.798 Krypton
37 Rb 85.468 Rubidium	38 Sr 87.62 Strontium	39 Y 88.906 Yttrium	40 Zr 91.224 Zirconium	41 Nb 92.906 Niobium	42 Mo 95.94 Molybdenum	43 Tc 98 Technetium	44 Ru 101.07 Ruthenium	45 Rh 102.906 Rhodium	46 Pd 106.42 Palladium	47 Ag 107.868 Silver	48 Cd 112.411 Cadmium	49 In 114.818 Indium	50 Sn 118.710 Tin	51 Sb 121.760 Antimony	52 Te 127.60 Tellurium	53 I 126.904 Iodine	54 Xe 131.294 Xenon
55 Cs 132.905 Caesium	56 Ba 137.327 Barium	57–71 Lanthanoids	72 Hf 178.49 Hafnium	73 Ta 180.948 Tantalum	74 W 183.84 Tungsten	75 Re 186.207 Rhenium	76 Os 190.23 Osmium	77 Ir 192.217 Iridium	78 Pt 195.084 Platinum	79 Au 196.967 Gold	80 Hg 200.59 Mercury	81 Tl 204.383 Thallium	82 Pb 207.2 Lead	83 Bi 208.980 Bismuth	84 Po 209 Polonium	85 At 210 Astatine	86 Rn 222 Radon
87 Fr 223 Francium	88 Ra 226 Radium	89–103 Actinoids	104 Rf 261 Rutherfordium	105 Db 262 Dubnium	106 Sg 266 Seaborgium	107 Bh 264 Bohrium	108 Hs 277 Hassium	109 Mt 268 Meitnerium	110 Ds 271 Darmstadtium	111 Rg 272 Roentgenium	112 Cp 277 Copernicium	113 Uut Ununtrium	114 Fl Flerovium	115 Uup Ununpentium	116 Lv Livermorium	117 Uus Ununseptium	118 Uuo Ununoctium

Lanthanoids (57–71)

57 La 138.905 Lanthanum	58 Ce 140.116 Cerium	59 Pr 140.908 Praseodymium	60 Nd 144.242 Neodymium	61 Pm 145 Promethium	62 Sm 150.36 Samarium	63 Eu 151.964 Europium	64 Gd 157.25 Gadolinium	65 Tb 158.925 Terbium	66 Dy 162.500 Dysprosium	67 Ho 164.930 Holmium	68 Er 167.259 Erbium	69 Tm 168.934 Thulium	70 Yb 173.04 Ytterbium	71 Lu 174.967 Lutetium

Actinoids (89–103)

89 Ac 227 Actinium	90 Th 232.038 Thorium	91 Pa 231.036 Protactinium	92 U 238.029 Uranium	93 Np 237 Neptunium	94 Pu 244 Plutonium	95 Am 243 Americium	96 Cm 247 Curium	97 Bk 247 Berkelium	98 Cf 251 Californium	99 Es 252 Einsteinium	100 Fm 257 Fermium	101 Md 258 Mendelevium	102 No 259 Nobelium	103 Lr 262 Lawrencium

Index